GENETIC ENGINEERING

A Reference Handbook
Second Edition

Other Titles in ABC-CLIO's
CONTEMPORARY
WORLD ISSUES
Series

Books in the Contemporary World Issues series address vital issues in today's society such as genetic engineering, pollution, and biodiversity. Written by professional writers, scholars, and nonacademic experts, these books are authoritative, clearly written, up-to-date, and objective. They provide a good starting point for research by high school and college students, scholars, and general readers as well as by legislators, businesspeople, activists, and others.

Each book, carefully organized and easy to use, contains an overview of the subject, a detailed chronology, biographical sketches, facts and data and/or documents and other primary-source material, a directory of organizations and agencies, annotated lists of print and nonprint resources, and an index.

Readers of books in the Contemporary World Issues series will find the information they need in order to have a better understanding of the social, political, environmental, and economic issues facing the world today.

GENETIC ENGINEERING

A Reference Handbook
Second Edition

Harry LeVine, III

**CONTEMPORARY
WORLD ISSUES**

A B C CLIO

Santa Barbara, California
Denver, Colorado
Oxford, England

Copyright 2006 by ABC-CLIO, Inc.

Library of Congress Cataloging-in-Publication Data

Le Vine, Harry.
    Genetic engineering : a reference handbook / Harry LeVine III. — 2nd ed.
        p.   cm. — (Contemporary world issues)
    Includes bibliographical references and index.
    ISBN 1-85109-860-7 (hardcover : alk. paper) — ISBN 1-85109-861-5 (ebook)
    1. Genetic engineering—Handbooks, manuals, etc.   I. Title.
II. Series.
    [DNLM: 1. Genetic Engineering—Handbooks.   QU 39 L433g 2006]
    TP248.6.L4 2006
    660.6′5—dc22

                          2006011074

09   08   07   06   10   9   8   7   6   5   4   3   2   1

This book is also available on the World Wide Web as an eBook. Visit abc-clio.com for details.

ABC-CLIO, Inc.
130 Cremona Drive, P.O. Box 1911
Santa Barbara, California 93116–1911

This book is printed on acid-free paper. ∞
Manufactured in the United States of America

Contents

Preface to the Second Edition

The second edition of *Genetic Engineering* covers accelerating advances in DNA technology and its applications since the first edition. The science continues to outstrip the ability of Western societies to fit the new capabilities with the philosophical and political expectations of their people and institutions. Many of the basic social and ethical issues raised at the outset remain. Legislative guidelines favored by most to provide basic protections of individual privacy and against genetic discrimination of various sorts are being established in many nations. The moral and ethical questions remain particularly divisive in the United States, reflecting the deep philosophical differences on many issues that polarize this nation.

Over the past six years, the nucleotide sequence of the entire human genome was determined; a furor over reproductive cloning sparked by the cloning of Dolly, a Finn Dorset lamb, crested and then subsided; and a new ethical controversy emerged—and is still growing—over the use of embryonic stem cells in regenerative medicine. The first generation of genetically modified crops is now established in world agriculture, and a second generation is in development that promises new questions about the role of agribusiness versus the traditional farmer. We are now able to assess the worldwide impact of early policies regulating genetically modified crops and reproductive cloning. Previously, there were no data, only sheer confidence—or fear. DNA testing, controversial at the O. J. Simpson trial, is now the gold standard for identifying individuals, and anthrax-laced envelopes revealed our vulnerability to bioterrorism.

The first chapter of the second edition provides an introduction to the science of genetic engineering, a history of the field, and a survey of the applications of genetic engineering. The new technologies driving progress are then described, including gene mapping, transgenic animals, gene chips, electronic information integration (or bioinformatics), the potential for personalized medicine (pharmacogenomics), and clonal stem cell technology for regenerative medicine. Chapters 2 and 3 discuss the problems, controversies, and solutions of the past and present from the U.S. and worldwide standpoint, respectively. Chapter 4 presents a chronology of discoveries that have led to genetic engineering and milestone accomplishments. Chapter 5 describes the lives and accomplishments of some of the contributors to the field. Chapter 6 is a collection of data and statistics about the applications and issues, and the last two chapters are a collection of resources for more information on topics covered in this book. A list of acronyms and a glossary of commonly used terms are also provided.

Preface to the First Edition

As nomadic peoples settled down, they brought useful plants under cultivation and gathered herds of animals. They learned to interbreed varieties to get larger, faster-growing stocks. These traditional means of genetic improvement culminated in the Green Revolution of the 1960s, in which high-yield and disease-resistant varieties of plants seemed capable of eliminating the specter of famine from much of the world.

In the early and middle 1970s, yet another revolution was brewing, one that had the potential to take genetic manipulation beyond people's wildest dreams. Human-mediated rearrangement of isolated pieces of genetic material comprising genes from different species, dubbed "recombinant DNA," was achieved in university laboratories. The awesome (or to some, awful) potential power of this genetic engineering caused scientists, in the fall of 1975 at Asilomar, California, to consider for the first time a self-imposed moratorium on certain types of experiments until the risks of the new technology were better understood.

The furor over regulation and application of genetic engineering continues today. Along with uncertainty over effects on the environment, application of genetic engineering emphasizes humankind's demonstrated lackluster competence in addressing social issues posed by new technologies, particularly before a crisis is reached. These range from heartfelt general questions of morality and ethics to privacy issues and access to information or fairness to different groups of people impacted by that information. Despite the often extreme positions taken by the media and various special interest groups, there is room on all of these issues for honest differences of opinion. No single point of view holds all of the "right" answers. The purpose of this book is to provide sufficient background on the issues involved with the application of genetic engineering to allow concerned citizens to participate in the decisions that must be made to fulfill the considerable promise of this new technology while avoiding both ecological disaster and the type of control over human life depicted in 1932 in Aldous Huxley's *Brave New World.*

The first two chapters of this book provide background material for understanding the technical basis of the science and a historical account of the evolution of genetic engineering and the issues surrounding its application. The third chapter sketches the lives of important contributors to the knowledge and the dialog. Chapter 4 collects data and documents and opinions central to the genetic engineering controversy. Chapters 5 through 7 provide a list of resources for those wishing to delve deeper into the subject, including organizations, print, nonprint, and electronic references. A glossary of genetic engineering terms provides explanations of the technical jargon.

1

Overview of
Genetic Engineering

What Is Genetic Engineering, and Why Is It Important Today?

In the early 1970s, a scientific experiment changed the relationship of humankind to the fundamental processes of nature. For the first time, DNA from one species of organism, *Xenopus laevis*, the African clawed frog, was purposefully transferred into another species, *Escherichia coli*, the common human intestinal bacterium. Nothing exciting happened. The bacteria grew normally, blithely replicating the piece of foreign DNA from another species inserted into a carrier plasmid of bacterial DNA in their cytoplasm. Although this experiment was in itself only an incremental extension of previous work, the workers in Stanley Cohen's and Herbert Boyer's laboratories had prepared the frog-bacteria plasmid in a test tube using isolated bacterial enzymes to cut and paste the DNA fragments together in a specific order. They had become genetic engineers, rearranging the DNA code for their own purpose. This simple demonstration ushered in the age of genetic engineering, where optimists foresaw that bacteria, yeast, plants, and animals could be modified to produce raw materials for industry, to improve food, to discover new medicines, to remove environmental contaminants, to recycle waste, and to provide permanent cures for inherited diseases.

But there was a cloud over this vision. To some people's minds, genetic engineering changed the world's natural order.

Ahead lay catastrophic disruption of the earth's ecosystem, uncontrolled spread of microorganism antibiotic resistance with attendant new plagues, and the corruption of the ideal of sanctity of life itself.

The Brief History of the Genetic Engineering Revolution

Immediately following these groundbreaking experiments came a call for a moratorium from both the lay public and some scientists to halt certain types of DNA transfer. This was an unprecedented turn of events in the scientific research community. During the summer of 1971, Robert Pollack at the Cold Spring Harbor Laboratories on Long Island convinced Paul Berg and others working on the monkey SV40 tumor virus to put off experiments transferring SV40 DNA, which was potentially tumorigenic to humans, into *E. coli*. Berg and Pollack subsequently organized the pivotal Conference on Biohazards in Cancer Research in January 1973 at Asilomar, California (Asilomar I), which was to set the tone for recombinant DNA research in the United States and much of the world.

The Asilomar Conference, sponsored by the National Science Foundation, the National Cancer Institute, and the American Cancer Society, all major government-funding agencies, was a rude awakening for the scientific community. For the first time, scientific policy was going to be decided, not through the traditional peer-review process by fellow scientists based on the scientific merit of the research, but by people who had a very different approach, lacked specific training in technical issues, and were pursuing a different agenda. The public feeling was that the potential impact of the new technology on society and the environment was too far-reaching for only scientists to decide what should be done. After all, some reasoned, what had scientists done with the knowledge of how to split the atom. A significant number of scientists were also worried. At the urging of Maxine Singer, the 1973 Gordon Research Conference on Nucleic Acids held a discussion on the moral and ethical issues of biohazards. A letter signed by all but 20 of the 142 Gordon conferees was sent to the president of the National Academy of Sciences and the president of the Institute of Medicine. It was eventually published in the trade journal *Science,* urging the establishment of a study committee to recommend specific guidelines "if the Academy and the Institute deemed it appropriate."

Reacting to the public concern and the mandate from the scientific community, the National Institutes of Health (NIH) formed the Recombinant DNA Molecule Program Advisory Committee in October 1974 and charged the group with framing guidelines to govern recombinant DNA research and with reviewing gene therapy protocols. Further discussions at the Asilomar Conference held in February 1975 (Asilomar II) led to a 16-month moratorium on recombinant DNA experiments until the NIH Guidelines became available in mid-1976. Strangely, human genetic engineering was specifically excluded from discussion at this time because it was considered too emotionally charged and was too far from realization at that point. Human engineering remains a sticky topic; witness the furor in February 1997 over the cloning of a sheep from a single adult udder cell in Scotland.

In the meantime, Senator Edward Kennedy and the Subcommittee on Health of the Committee on Labor and Public Welfare began the first public debate on recombinant DNA in April 1975. This discussion was echoed at the local level as communities such as Cambridge, Massachusetts, with industrial or academic recombinant DNA practitioners held town meetings seeking to regulate the technology in their jurisdiction. Only concerted public relations efforts and information exchanges between the universities and the public averted a fear-driven shutdown of the research institutions. As a result of the early concern, sixteen bills were introduced in Congress to regulate recombinant DNA research; none were passed into law. A U.S. National Academy of Sciences forum on industrial applications of recombinant DNA technology held in Washington, D.C., March 7–9, 1977, was turned from panel discussions into a debate and media event. Dissenters in the audience disrupted the proceedings in Vietnam War protest style. Nevertheless, by that time, it was becoming apparent that the doomsday scenarios had been exaggerated and that working with the technology under the NIH Guidelines was generally safe. In the ensuing years. the restrictions of the guidelines were gradually relaxed as data accumulated on safety concerns, suggesting that the technology could be controlled.

In October 1990, the Department of Energy began the Human Genome Project, a massive endeavor to determine the nucleotide sequence of the genome, all 3 billion nucleotides of the DNA in the 23 human chromosomes, and to put into place the technologies required to use that information in scientifically, medically, and ethically responsible ways. The project had seven major goals: 1) map and sequence the human genome with an emphasis on identifying genes; 2) map and sequence the genomes of five model laboratory organisms—the laboratory mouse, the bacterium *Escherichia coli*, the roundworm *Caenorhabditis elegans*, and the fruit fly *Drosophila melanogaster*; 3) identify social, legal, and societal issues to anticipate and plan for problems (Task Force on Ethical, Legal, and Social Implications); 4) develop information and analysis systems to allow the genome project information to be used worldwide by researchers; 5) improvement of technologies for genome study such as DNA sequencing; 6) support transfer of genome project technology to industry and other areas where it might be useful; and 7) support training of students and scientists in the various skills needed for genome research. The need for genome sequencing was hotly debated. Critics opposed the draining of resources away from other, more creative science; questioned the wisdom of determining DNA sequence, 98% of which does not appear to encode a genetic message; and finally, asked *whose* DNA would be sequenced. Most people, however, believed that locating the human counterpart of a protein for which the function had been determined in other animal systems would enormously advance scientific and medical understanding. Such understanding was expected to lead to effective therapies, thereby saving lives and reducing suffering.

The International Human Genome Sequencing Consortium published a first-draft, 90%-complete sequence of the human genome in the February 15, 2001, issue of *Nature*, which was followed up by the complete sequence on April 14, 2003. The big surprise was that the human genome had a smaller number of genes than expected—indeed, it had fewer than some "lower" organisms, 30,000–40,000. The smaller number of genes was compensated for, however, by more complex processing to give more versions of each gene product.

Although the apprehension over handling recombinant DNA technology eased with the increased knowledge about safety and with familiarity, concern over the impact of applications of the technology continued to build. There remains substantial controversy over the ecological impact of widespread dissemination of genetically modified organisms. What may very well turn out to be the real hazard, however, is the social impact of what can be done with genetic engineering technology. Debate has shifted from the dangers of the technology itself to what societies will do with the information and the ability to manipulate genes. The consequences of the economic changes wrought by the new industries born of genetic engineering on developing countries are yet another area of concern. What was once a scientific problem is now a social one. How will we use and control the use of our newfound abilities? It is unsettling to know that it is entirely in our hands to make the world better—or worse.

Nuts and Bolts of Genetic Engineering

Many scientists contributed to developing the ideas and methods crucial for making recombinant DNA a useful technology. Some of those providing the most insightful ideas were honored by the awarding of the Nobel Prize to recognize the accomplishment. The number of winners whose contributions benefited genetic engineering (Table 1.1) testifies to the development in scientific understanding required for the technology to exist. As with all good explanations, it turns out that the ideas required to understand

TABLE 1.1
Nobel Prize Winners Contributing to Genetic Engineering

1905	Robert Koch	M&P	Elucidation of pathology of tuberculosis and principles of culture of microorganisms
1910	Albrecht Kossel	M&P	Studies on chemistry of the cell distinguishing proteins and nucleic acids
1915	Sir William Henry Bragg	Physics	Analysis of structure by X-ray crystallography
1915	Sir William Lawrence Bragg	Physics	Analysis of structure by X-ray crystallography
1933	Thomas Hunt Morgan	M&P	Role of chromosome in heredity
1958	George Wells Beadle	M&P	Gene control of cellular chemical synthesis
1958	Edward Lawrie Tatum	M&P	Gene control of cellular chemical synthesis and genetic recombination
1958	Joshua Lederberg	M&P	Sexual transfer of genes between bacteria, leading to early genetic engineering
1958	Frederick Sanger	Chem	Structure of proteins—insulin
1959	Severo Ochoa	M&P	Biosynthesis of RNA and DNA
1959	Arthur Kornberg	M&P	DNA polymerase and DNA synthesis
1962	Francis Harry Compton Crick	M&P	Structure of DNA, genetic code
1962	James Dewey Watson	M&P	Structure of DNA, viral structure, protein biosynthesis
1962	Maurice Hugh Frederick Wilkins	M&P	Structure of DNA

continues

TABLE 1.1 continued

1965	Jacques Lucien Monod	M&P	Mechanisms by which genes are regulated and proteins manufactured
1965	Francois Jacob	M&P	Action of regulator genes, bacterial genetics
1965	Andre Michael Lwoff	M&P	Replication and genetics of viruses and bacteria
1968	Robert William Holley	M&P	structure of nucleic acids, sequence of phenylalanine tRNA
1968	Har Gobind Khorana	M&P	Synthesis of polynucleotides, the genetic code
1968	Marshall Warren Nirenberg	M&P	Method for deciphering genetic code, determining protein amino acid sequence from DNA
1969	Max Delbrück	M&P	Genetics of bacteriophage recombination
1969	Alfred Day Hershey	M&P	Replication, genetics, and mutation of bacteriophages
1969	Salvador Edward Luria	M&P	Replication, genetics, and mutation of bacteriophages
1972	Christian Boehmer Anfinsen	Chem	Control of protein folding by amino acid sequence
1972	Stanford Moore	Chem	Automatic amino acid analyzer and the sequence of ribonuclease
1972	William Howard Stein	Chem	Automatic amino acid analyzer and the sequence of ribonuclease
1975	Howard Martin Temin	M&P	Interaction between tumor viruses and cellular genetic material, and reverse transcriptase
1975	Renato Dulbecco	M&P	Molecular biology of tumor viruses
1975	David Baltimore	M&P	Interaction between tumor viruses and cellular genetic material, and reverse transcriptase
1977	Rosalyn Yalow	M&P	Radioimmunoassay technique
1978	Daniel Nathans	M&P	Development of restriction endonucleases
1978	Hamilton Othanel Smith	M&P	Development of restriction endonucleases
1978	Werner Arber	M&P	Development of restriction endonucleases
1980	Paul Berg	Chem	Biochemistry of nucleic acids
1980	Walter Gilbert	Chem	Method for DNA sequencing
1980	Frederick Sanger	Chem	Method for DNA sequencing
1982	Sir Aaron Klug	Chem	Electron microscopy of nucleic acids
1983	Barbara McClintock	M&P	Chromosomal exchange of genetic information and mobile genetic elements
1984	Robert Bruce Merrifield	M&P	Chemical synthesis on a solid support
1984	Niels K. Jerne	M&P	Specificity in the immune system and principles for monoclonal antibodies
1984	Georges J. F. Köhler	M&P	Specificity in the immune system and principles for monoclonal antibodies
1984	César Milstein	M&P	Specificity in the immune system and principles for monoclonal antibodies
1990	Sidney Altman	Chem	Discovery of catalytic RNA — "ribozymes"
1990	Thomas R. Cech	Chem	discovery of catalytic RNA — "ribozymes"
1993	Kerry B. Mullis	Chem	Invention of polymerase chain reaction to amplify DNA
1993	Michael Smith	Chem	Oligonucleotide-based mutagenesis of DNA

M&P = medicine and physiology; Chem = chemistry.

the basic principles of molecular biology and genetic engineering are elegantly simple in concept if not in practice to utilize.

The Structure of DNA

Recombinant DNA technology is the science of handling and manipulating the genetic material of cells. The word *recombinant* means "new combinations" and refers to the shifting of genetic material from one organism into another of either the same or a different species. Nature can also move genetic information from one cell into another by viruses or by "jumping genes," first described by Barbara McClintock in 1951, but it happens (fortunately) only rarely under normal circumstances. Scientists called molecular biologists have learned how to speed up and to control the transfer process as well as how to transfer DNA between species. The genetic material is made of a chemical polymer called deoxyribonucleic acid, or DNA. The DNA polymer consists of similar types of units connected end to end like a string of beads, with each bead representing a deoxyribonucleic acid unit. DNA is very long and thin. Stretched out, the DNA contained in a human cell would be 6 feet long, but it is so thin that 500 pieces laid side by side could pass through the eye of a sewing needle. A champion for DNA is the lungfish. Its cells contain DNA that could stretch 1,138 feet, almost two-tenths of a mile, wrapped up in a region of the cell called a nucleus only about a hundred thousandth of an inch across. A regular microscope can't see a strand of DNA; it's too thin. An electron microscope magnifying nearly a million times is needed to see a DNA strand clearly. To fit into the nucleus of a human skin cell one-hundredth the size of a grain of rice, the DNA is wound tightly to form chromosomes. There are 23 pairs of chromosomes in the nucleus of normal human cells. They become visible in a regular light microscope when the cell copies its genome as it prepares to divide into two daughter cells.

The chemical units strung together in a DNA strand are even tinier, so small that the most powerful electron microscope can't make them visible. These beadlike units come in four "flavors," designated by the letters A, C, G, and T, which represent the nucleic acid bases adenine, cytidine, guanine, and thymine, respectively. The backbone of each strand is formed by a phosphate ester link between the 3' and 5' positions of successive sugar residues. The two single strands of DNA are wound in a helical arrangement around each other as shown in Figure 1.1.

FIGURE 1.1
The DNA Helix

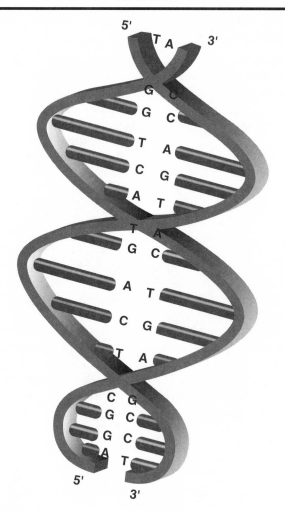

The bases form a hydrogen-bonded core, like rungs on a ladder, with the backbone phosphates facing the outside similar to the rails of the ladder. Twisting the ladder lengthwise forms a helix. The letters signifying all of the bases in the DNA of a human cell would fill a 1,000-volume encyclopedia. It is the sequence of these units, read left to right (or 5-prime to 3-prime along the deoxyribose sugar-phosphate backbone, 5' to 3', in molecular biologese) that is important. Groupings of these strings are the DNA code, instructing a restriction enzyme to cut the DNA at one place or a protein to bind to the DNA somewhere else. Three-letter arrangements (codons) code for proteins that form the cell and manufacture its chemical parts.

Having two of these DNA strands wound around each other in the cell helps to prevent cells from erring when making new DNA for a daughter cell. New DNA is always copied from old DNA, which accounts for how information about hair or eye color or blood type is accurately passed from parent to child. The intertwined DNA strands of the chromosome are "complementary" copies of the same code. Where one strand has an A, the other has a T, and where one strand has a C the opposite has a G. A-T's and C-G's pair up, sharing hydrogen bonds between the strands and holding them together (see Figure 1.1).

The DNA Code

Although at first glance the DNA sequence seems random, in fact the DNA units are grouped in "words," the length of which depends on the meaning. Sometimes the words overlap so the message depends on where you start reading. Four-, five-, six-letter, and longer words are read by parts of the cell that control what the DNA sequence is being used for. Some of the longer words specify binding of certain proteins to that site such as a transcription factor binding to start messenger RNA (mRNA) production by an RNA polymerase or for binding of DNA polymerase to start DNA replication (Figure 1.2).

FIGURE 1.2
Recognition Sites for Proteins on DNA

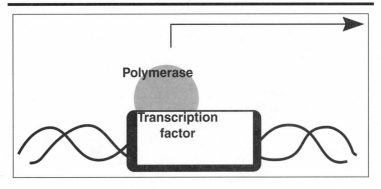

There are three-letter words for amino acids, which are chemicals found in cells. DNA words are grouped into "sentences" called genes, which specify one protein molecule. Clever experimentation eventually deciphered the code. These special words are first copied in sequence from the DNA into another chain, this time made of ribonucleic acid units that are strung together similarly to DNA called messenger RNA (mRNA).

A cell often makes many copies of a particular RNA message, unlike DNA. This message is disposable and is destroyed when it isn't needed anymore. The message instructs another part of the cell, known as a ribosome, to join the 20 varieties of an amino acid "bead" in a specified order end to end into polymers to make proteins. There are thousands of ribosomes in a cell, on which many different proteins are being made inside a cell at any one time. Proteins conduct the business of life in a cell. Although the DNA is like the card or computer catalog of a library forming a list of all of the books therein, proteins are like the people who use the library. In multicellular organisms each cell contains all of the genetic information required to specify a whole organism. Only a small part of the DNA of a cell is in use at any one time, with the particular part depending on the type of cell. The unused DNA is wound up tightly out of the way on spool-like structures and twisted onto the scaffold of the chromosome.

Proteins act as catalysts and building blocks to manufacture the other components of a cell, including carbohydrates (sugars) for energy, the cellular protein skeleton, and lipids for membranes as well as the DNA and RNA. Protein catalysts known as enzymes also construct small molecule metabolites from chemi-

FIGURE 1.3
Coding for Protein Synthesis and DNA Replication

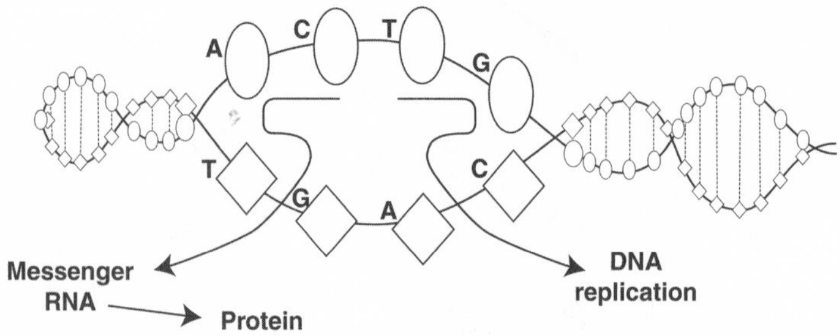

cals in their environment. These are intermediates that transfer needed energy and chemical groups within the cell and help the cell survive and carry out its function. In industrial applications with bacteria or fungi, some of these metabolites are useful to humans as food (sugars, vinegar), for manufacturing (ethylene glycol, polymers), or as medicines (antibiotics).

All animal, plant, and microbe cells use universal code words for the amino acids. Thus, the same protein will be made from the same messenger RNA code in all living things. This is what makes biotechnology work. Whatever type of cell makes the protein, bacterial, fungal, plant, snake, or human, it will be the same. It is the order or sequence of the amino acid beads in a protein's polypeptide chain that sets the shape of the protein and determines what it does.

Cloning Technology

Recombinant DNA technology allows the DNA sequence information coding for making a protein (called an insert) such as the enzyme adenosine deaminase (or ADA) to be placed into a small circular piece of DNA called a plasmid vector. The process is depicted in Figure 1.4.

The plasmid and insert are treated with the same restriction endonuclease, such as BamHI, which generates complementary or "sticky" ends by hydrolyzing the phosphate ester backbone connecting the nucleotides at specific sites (arrowhead). The endonuclease recognizes the nucleotide sequence GGATCC and cuts between the GG pair on the 5' end of each strand. After mixing the insert and vector, a DNA-joining enzyme, DNA ligase, is added to reconnect the phosphate ester backbone of the helix. A cell (microbial or a cell from a multicellular organism) is then made to take in the vector (transformed) with the DNA, which now codes for the ADA protein. The vector plasmid with its ADA gene is copied when the cell multiplies so that every cell has a copy of the ADA information. The vector contains DNA sequences that direct the host cell's machinery to make mRNA from the insert sequence, which is then translated into the actual ADA protein in the cell. This technique is routinely used to produce a protein in a laboratory for study. It can also be used to provide gene therapy to replace a deficient gene product in a seriously ill patient. W. French Anderson and his fellow medical scientists used just such a trick to repair the white blood cells of

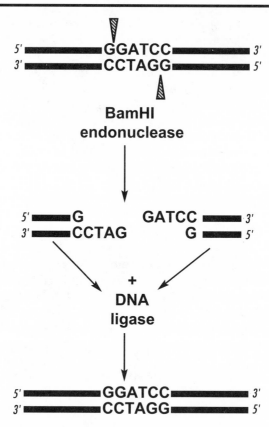

FIGURE 1.4
DNA Cloning with the BamHI Restriction Endonuclease

two children who were always getting infections because their white blood cells could not make the important ADA protein.

Polymerase Chain Reaction

In addition to the enzymes and vectors used to manipulate DNA in a test tube, a technique called polymerase chain reaction, or PCR, is used to make an unlimited number of copies of a chosen DNA sequence. This technique won a Nobel Prize for its originator, Kary Mullis, in 1993 and launched another revolution within the recombinant DNA revolution. The principle behind the technique is stunningly simple, yet immensely powerful. PCR selectively copies a single DNA sequence from among a mixture of millions of DNA sequences and amplifies it for further use. The simplicity comes from the use of a heat-stable DNA polymerase from a microorganism that lives near the boiling point of water in hot springs and near superheated undersea volcanic vents. This enzyme catalyzes multiple rounds of replication of the chosen DNA template.

In PCR, short sequences (15–25 nucleotides) called *primers* are supplied by the investigator as chemically synthesized short nucleotide sequences complementary to (matches) the two ends of a DNA sequence (template) that is to be copied. The choice of these primer sequences determines the part of the whole DNA sequence to be copied by serving as the starting and ending points. After waiting long enough for active polymerase to copy the DNA sequence starting at the primer-DNA matched sequences, the mixture is heated to separate the new DNA strands from the old and to inactivate the polymerase temporarily. After cooling, the cycle is repeated, more primers bind, and the reactivated polymerase makes another copy of the DNA. After the first few rounds, most of the target DNA sequences in the sample are PCR copies of the original sequence bounded by the primers, and this number increases exponentially with the number of copy cycles. After 32 cycles, a typical amplification, the number of copies of the target DNA sequence have increased 2^{32}-fold or more than 4 billion-fold!

PCR's enormous capability to amplify specific sequences from among many others in a sample is particularly useful in forensics where DNA samples from a crime scene are frequently limited. The copies are subjected to analysis by the standard forensic DNA methods. PCR can also be used to engineer cloned sequences precisely instead of having to rely on finding restriction endonuclease sites nearby. Limited primarily by the ingenuity of the user, modified PCR techniques can create changes to insert or delete DNA sequences from a target sequence or measure levels of an RNA message in cells.

Chip Technologies

The Human Genome Project stimulated the application of high-throughput, high-density technologies in use in the microelectronics industry to the analysis of genes and their nucleic acid products. These are known as chip technologies because they are generally array-based and printed on a microscale on glass slides or chips. A small number of chip arrays (and soon a single chip array) contain complementary sequences for each of the 30,000 or so genes of the human genome. Based on the ability of nucleic acid sequences to recognize complementary nucleic acid sequences and coupled to PCR amplification methods, the utility of the chips depends on the ingenuity of designers to formulate questions. They can also be used to deduce the sequence of a stretch of DNA or search for marker sequences in gene mapping projects. By measuring messenger RNA levels, they can determine patient responses to a therapeutic regimen (pharmacogenomics) and assess prognostic indicators for a patient (Friend and Stoughton 2002). As a diagnostic, they can rapidly identify microorganisms responsible for an infection using chips containing specific pathogen-derived DNA sequences.

Information Management and the Rise of Bioinformatics

The immense scope of the sequencing effort of the Human Genome Project fostered the development of new technologies. They needed to generate the raw sequence data, to process, to check for errors, to align the sequences, and to assign them to the 23 human chromosomes. An integral part of the Genome Project was the development of sophisticated data management systems to process the sequences and to put the final data in a form that would be accessible to many types of users. The Genome Project ushered in the serious application of bioinformatics, which is discussed separately because it has changed the way biomolecular science is done.

Once the human genome sequencing was finished other genomes were moved up in the list for sequencing. For the human genome, however, the focus changed to information management. Because only 2% of the human (and other mammals) genome sequence codes for functional genes, connecting the DNA sequence data with identified genes was quite a chore. Suspecting this, the sequencers started with gene-rich regions and filled in the other parts later. Called "junk DNA" by some, the history of the evolution of the human genome is recorded in this noncoding DNA because those portions were never expressed to be selected for or against.

In the database, the DNA sequence is laid out along the linear chromosomes with the positions of known genes and the regions of chromosomes associated with genetic diseases or disease risk mapped onto the sequence. The mandate for the project was that all of the incredible amount of information was to be made widely available. The World Wide Web was the answer to disseminating the information to anyone who wanted it. You can access the human genome sequence as well as the genomes of an increasing number of organisms at http://www3.ncbi.nlm.nih.gov, as well as a variety of educational opportunities. Much of that information is annotated and connected with other databases with information about function, linkage to genetic diseases, and information of use to the scientists who use the genome sequence in their work. The genomes of agriculturally and industrially important plants, animals, and microorganisms, as well as those of medical importance, are also available from that Web site. Key animal models for research such as *Drosophila melanogaster* (fruit fly) and the zebrafish have their own systems with their specialized data.

The determination of the genes responsible for diseases has been improved by the availability of the genome sequence. Natural variation, called *polymorphisms,* among individuals in the DNA sequences in between genes provides a large number of closely spaced markers for locating candidate disease genes or genes that contribute to the risk of a disease. Comparison of affected and control groups of individuals through a complex process called statistical genetics associates the disease with particular areas of chromosomes marked by specific polymorphisms. Nearby genes are candidates for the cause of disease or increased risk for developing the disease. Candidate genes can become targets for drug discovery for the development of therapeutics. A good polymorphic marker or multiple markers for the risk of developing a disease are useful for counseling even if the gene is unknown or the way it increases risk is not understood. Someone with a marker associated with increased risk for hypertension can be advised to make lifestyle changes, such as lowering salt intake, even before symptoms appear to avoid severe problems later in life.

Gene Mapping

Microfabrication techniques similar to those used to create electronic memory chips for computers have been used to attach the probes for specific DNA sequences in known positions at high densities onto special surfaces. All of the genes of the human genome—more than 30,000 of them—are represented on a single or small number of chips. Similarly, tens of thousands of probes for markers called polymorphisms at different positions on chromosomes can be attached to chips, and the presence or absence of all those markers in an individual can be tested at one time. Geneticists can follow the segments of chromosomes as segments assort through the generations in a family, which helps investigators to associate particular traits or a disease with the inheritance of portions of a chromosome. In this way, they can find disease genes as well as risk and protective factors when the gene responsible is eventually identified.

Pharmacogenomics

From a seemingly impersonal high-throughput procedure arises the possibility of truly personalized therapies. Already people with certain genetic forms of enzymes that metabolize some drugs and not others benefit from a diagnostic screening and their medication selected to provide the maximum therapeutic benefit with minimal side effects. This "profiling" has the potential to be extended to selecting therapeutics based on particular groupings of polymorphisms that have been associated with the best response to a particular treatment. This ultimate in personalized medicine remains mostly in the future while more data are collected and issues of medical privacy are settled.

Although the genome indicates potential, it does not indicate when, if ever, a feature will be expressed, or if it is expressed, to what extent. It reflects "the odds"—a likelihood rather than a certainty. Closer to true manifestation of a trait is the actual transcription of the gene linked to that trait into the messenger RNA that codes for the protein that when translated will produce the protein itself, which can then have its effects. Messenger RNA expression can also be subjected to analysis to determine responses to treatments. The levels of mRNA transcribed from each of the genes of interest are measured with chips bearing complementary sequences to the mRNA. The expression levels are scrutinized for patterns that correlate with the prognosis for breast cancer, for example (Friend and Stoughton 2002), or with the efficacy of a treatment. This provides a preview of the efficacy of a treatment in time to adjust the regimen without having to wait for symptom relief. This is particularly valuable for conditions that change slowly.

Bioinformatics

The sheer amount of sequence data in the human genome is beyond analysis by the unaided human mind. Add to this the sequences of the multiple alternative forms of the messenger RNAs that code for proteins, and the information overload is incomprehensible. The sequences of other organisms are constantly being added (for an update, see http://www3.ncbi.nlm.nih.gov). Computers and special software are now indispensable both for finding sequences of interest and in defining what other sequences are interesting. The analysis of patterns in DNA sequences, much like patterns in speech or handwriting, gives a hint as to organizing principles that seem lost in the details. In silico molecular biology, where the scientist never touches the DNA or a test tube, deduces the logic of the organization of genes and the chromosomes of which they are part.

Genome organizational hypotheses required the development of sophisticated computational analysis. Frequently, the programs are written by nonbiologists who are skilled at teasing order out of apparent chaos by dealing with the properties of information that are not necessarily related to the specific topic. For information specialists and their programs, reading the sequence of the genome is similar to trying to break a secret code. Just like the English alphabet, the four-"letter" DNA alphabet—adenosine, thymidine, cytosine, and guanosine (ATCG)—is formed into words, phrases, even paragraphs. If the reader doesn't know the language's rules about how letters, words, and phrases are put together, the strings of characters are gibberish.

Computer analysis of the number of times a letter combination is used or what letters most frequently appear next to one another gives hints about the way that a message is coded. For example, in the human genome, an ATG word, or codon, specifies the incorporation of the amino acid methionine into a protein. It can signal the beginning of a protein, because proteins always start with a methionine, or it could be a methionine somewhere else in the protein. True starting methionine codons, however, have been noted to be located near a CAAT sequence. Some of the amino acids are specified by more than one codon triplet, and thus there are multiple "spellings" for some amino acids. This redundancy comes about because there are 64 ways to arrange four letters, taking three at a time, but only 20 amino acids.

The "spelling" or codon usage is different inside genes than outside of them, another pattern that can be used to recognize genes. This is important for picking out human genes because in the higher organisms such as mammals only about 2% of the genome DNA sequence codes for genes. Additional computer analyses tell much more about the genome, but they are beyond the scope of this book.

The leap of faith into believing computer analyses and statistical arguments was difficult for molecular biologists, who as experimentalists needed to be shown the relevance and usefulness of these apparently unrelated techniques. Much like the quantum theory in physics, informatics is now accepted by practitioners of molecular biology because it predicts and explains many phenomena.

Genetically Modified Animals and Plants

Although the details of genetically modified organism generation are beyond the scope of this book, a number of unifying concepts will help in thinking about them. It is also important to be aware of the differences, both perceived and real, between transgenic organisms, that is, organisms derived by nuclear transplantation (nuclear cloning), and organisms or tissues derived from stem cells of various sorts.

Transgenic organisms incorporate a gene from another species into reproductive cells, the germ line, and thus into all cells of that organism and its progeny. For plants that replicate vegetatively (without sexual reproduction—seeds, pollen, or spores), all cells contain the new gene. The transgene may be another species' version of the gene in the recipient, or it may be a gene for which there is no counterpart. By this definition, bacteria into which nonbacterial genes have been inserted are also transgenic; however, this term generally refers to multicellular organisms into which foreign genes have been incorporated.

The most common transgenic animal species is the mouse. To create a transgenic mouse, the gene of interest is isolated from cells derived from the donor species using techniques described earlier for the manipulation of DNA. The gene is incorporated into a plasmid that also contains a promoter DNA sequence that will determine how the expression of the transgene is regulated in the mouse. The specific details vary depending on the strategy for generating the transgenic animal. In the most common method, the construct containing the cloned gene with its selected promoter is then injected into the early-stage nucleus of a fertilized ovum where it incorporates into the DNA of one of the mouse chromosomes. This modified ovum is implanted in the uterus of a hormone-treated pseudo-pregnant surrogate mother where, if all goes well, it will develop into a mouse. All of the cells of the mouse embryo that develop from this transgenic ovum will contain the new gene. The gene will be expressed under the direction of its attached promoter, which may be different than the genes around it, which are controlled by the cell type such as a lung cell or liver cell.

Another method modifies mouse embryonic stem cells (ES cells) in isolated cell culture with the foreign gene and then implants early-stage embryos derived from individual stem cells in pseudo-pregnant mothers as before. An advantage to the ES cell technique is that the cells can be selected for their expression of the new gene before they are implanted. Screening cells early on reduces the number of offspring that have to be produced to obtain pups positive for expression of the transgene.

Nuclear Transplantation

The most controversial form of genetic manipulation burst onto the scene in 1997. Dolly, a Finn Dorset sheep, was produced by the transplantation of the nucleus from an udder mammary tissue cell of one sheep into an ovum from which the nucleus had been removed. Nuclear transplantation for human cloning has proved to be a "hot-button" controversial issue. Because the genes on the chromosomes of every cell in the body are the same, the nucleus from a donor cell theoretically can be derived from anywhere in the body. The nucleus is removed from the donor cell with a micropipet and injected into an unfertilized host ovum whose own nucleus has been removed. In a key series of events including treatment with biological factors, the donor nucleus de-differentiates from the type of cell it came from and takes on the characteristics of the ovum. As a germ cell, the ovum gives rise to cells that differentiate into all of the different kinds of cells in the body. Genetically, the cell containing the transplanted nucleus differs from the donor only in the small amount of genetic material carried by the mitochondria in the cytoplasm of the ovum. They are genetically like a normal biological identical (monozygotic) twin, more closely related to the nucleus donor than fraternal twins are to each other. Biological monozygotic twins arise from a splitting of an embryo at an early stage in utero for an unknown reason.

Although a number of species have been successfully "cloned" by nuclear transplantation, the success rate (number of viable offspring per ovum injected) for nuclear transplantation is presently very low and depends on the species. The offspring frequently develop problems as they grow up that have been attributed to incomplete de-differentiation of the donor nucleus which disrupts some of the developmental instructions. Dolly was euthanized in February 2003 after progressive lung disease was diagnosed. She had suffered from several ailments suggestive of premature aging. Even if these technical issues are resolved, the debate over creating new individuals is inseparable from the strong sentiments and ethical implications that surround anything that involves human reproduction.

Stem Cell Regenerative Therapy

In November 1998, two research groups including the Geron Corporation (Menlo Park, California) announced that they had isolated and maintained human embryonic stem cells in culture. Unlike most work of this type, Geron had been discussing the ethical dimensions of the work they were undertaking with the Graduate Theological Union (Berkeley, California) beginning in late 1996. The Clinton administration published federal guidelines for funding of stem cell research in August 2000, leaving the details to the incoming George W. Bush administration. On August 9, 2001, President George W. Bush announced the decision allowing federal funding of the use of existing pluripotent stem cell lines derived from human embryos prior to that date.

Human embryonic stem cell lines were a major development because these cells have the ability to develop into most of the specialized cells and tissues of the body. Although mouse embryonic stem cells had been cultured since 1981, translating that success into the human system had been elusive. To distinguish nuclear transplantation that would result in a new individual (reproductive cloning) from that involving embryonic stem cells (regenerative technology), the stem cell process was dubbed therapeutic cloning. Because embryonic stem cells were obtained from human embryos that were destroyed in the process, the reproductive rights and abortion issues ignited immediately. Stem cell technology is focused on cell and tissue replacement rather than creation of a new organism, and thus this controversy was spared that set of ethical issues.

Small groups of nondescript stem cells dwell in the marrow of the long bones and in the sinuses of the thymus. They await hormonal signals to trigger them to divide and turn on the genes that remodel them into the next generation of red blood cells, white blood cells, and platelets. There is a continuum of stem cell types, each more specialized than embryonic stem cells. Failure to recognize the multiplicity of kinds of stem cells confuses discussions in the lay media. Only stem cells originating from embryos or possibly those derived from a fetus, depending on how they were obtained, run afoul of the ethical issues associated with embryonic stem cells. Hematopoietic (blood-forming) stem cells are adult stem cells that will only give rise to a highly restricted lineage. There are, in addition, a variety of precursors of specialized cells poised in various areas of the body ready to be awakened to their particular task and to integrate into specific pathways. Umbilical cord blood is a source of early precursors although these stem cells are of later stages of commitment than the embryonic stem cells. These precursors are usually only able to give rise to a restricted range of cell types. They are considered pluripotent as opposed to the totipotency of early embryonic cells or the ovum, which have the capacity to form any cell in an organism.

Stem cells can be therapeutically useful. In cases of certain white blood cell cancers, the endogenous blood cell precursors in the marrow are killed by radiation and by chemotherapy. The marrow is then repopulated by transplanting noncancerous marrow containing marrow stem cells from a donor with a matching tissue type (analogous to blood type). The stem cells take up residence and resume producing new blood cells for the host. Stem cells from other tissues are treated with an appropriate growth factor cocktail and transplanted into the area they are supposed to colonize.

Although all cells in the body are ultimately descended from the single fertilized ovum, embryonic development proceeds along lineages in which certain kinds of cells are derived from one particular group of cells and not from others. In simpler organisms such as the nematode *Caenorabiditis elegans,* the lineage and fate of every single cell has been traced. As development continues building the organism, cells become more specialized and are locked into fixed roles and their genetic machinery adjusted to maintain that state. Earlier-stage stem cells, cells that are not yet committed far down the pathway, are more abundant in embryonic tissues than in adult tissue. Depending on the conditions and the sequence of signals, embryonic stem cells can be induced to differentiate along selected lineages. In the right environment, which includes the appropriate cell contacts and secreted materials as yet unknown, a cell can integrate into a tissue or join with other cells to form a new tissue and hopefully adopt its function. The implications of this procedure for biomaterial production or tissue regeneration are obvious and profound. However, the use of human embryonic tissue, no matter how it is derived, as a source of stem cells runs afoul of ethical and religious concerns. This "hot-button" status and governmental reaction have strictly regulated stem cell research in the United States. Other countries are less restricted by these attitudes.

There are many difficulties in identifying and isolating the small numbers of adult stem cells and then in expanding the cell population to usable numbers without compromising their ability to develop into multiple cell types. These difficulties are reduced by isolation of stem cells from earlier developmental stages. Fetal stem cells isolated from early-term fetuses, in the case of humans from miscarried or aborted fetuses, are present in larger numbers and have a broader tissue differentiation spectrum than adult stem cells. Finally, totipotent embryonic cells can be isolated from the unorganized inner cell mass of the embryo at the stage the outer layer of cells that will become part of the placenta separates from the cell mass. Finally, embryonic germ cells can be isolated from the germinal ridge of the developing embryo at a slightly later stage. These cells were destined to become the reproductive cells and thus are also totipotent.

Embryonic stem cells, in addition to being capable of differentiating into any kind of cell given the proper clues, are also truly immortal in that they will continue to divide, producing identical daughter cells. Later stage stem cells can also be cultivated and expanded in culture, but like differentiated adult cells, they can only divide a restricted number of times. Embryonic stem cells are clearly the most useful type of stem cell, but they come at the cost of the destruction of the embryo from which they derive. Like nuclear transplantation (reproductive cloning), stem cell research is a divisive issue. It pits patients with deadly or debilitating diseases or conditions with no other hope for relief against those who feel that they are protecting the unborn and the sanctity of human life from exploitation.

Stem cell technology is not designed to re-create an individual, but to generate differentiated cell types for transplantation to replace damaged or defective nerves, heart, kidney or other cells. It does not involve genetic manipulation. Conceivably, new stem cell techniques could induce more differentiated stem cells from the individual needing a transplant to de-differentiate and take on the required characteristics. The details of the technology surrounding the successful functional incorporation of differentiated stem cells remain to be worked out, particularly for the human system. Research in this area in the United States is currently limited by restriction of federal government funding of human stem cell research. Although work is continuing with mice and other animals, there are significant differences in development between humans and other animals, and some of the steps cannot be modeled in rodents. The technology has the potential to incorporate genetic manipulation, although at present, that is not at issue.

Applications of Genetic Engineering

Recombinant DNA technology is a variety of enzymatic and chemical procedures used to manipulate DNA in the test tube to form selected combinations of sequences. By these techniques, genes can be added to or removed from the genome of a cell, or existing genes can be modified in some way. The process of changing the genetic complement of a cell or a whole organism by this process has come to be known as genetic engineering. The process of rearranging DNA sequences in vitro is key in this definition, because accomplishing similar rearrangements by traditional breeding practices is associated with different regulatory guidelines. The technology is rapid and powerful, permitting much more far-reaching human control over the biological world. It is also more precise, capable of controlling single genes. Since the advent of the recombinant techniques in the early 1970s, ever more applications of genetic control have been demonstrated in the areas of health, agriculture, industrial production, and environmental remediation. It has opened up new ways of doing things biologically that had been done previously chemically, if they could be done at all.

Many kinds of host cells from various organisms can be used to make useful recombinant proteins from DNA sequence encoded in vectors. Very often these hosts are simple bacteria or fungi that can be grown easily in thousands of liters of culture media. Not only single-celled organisms but plants or animals made up of many cells such as soybeans and mice and cows can also be caused to make a chosen protein. They are known as transgenic because they carry a gene from another kind of organism, a *trans*gene. Soybeans harboring a gene making them resistant to weed killers or a protein toxic only to insects can make farming them easier. Expressing an altered human gene for the protein superoxide dismutase in a mouse produces a form of Lou Gehrig's disease (amyelotrophic lateral sclerosis) in the mouse, which can be used to help researchers find a cure for that neurological disease. Transgenic animals can produce useful human biological products (Velander, Lubon, and Drohan 1997). The milk from transgenic cows in which particular cow protein genes are replaced with the human protein sequence is a valuable product, an infant formula for babies allergic to cow's milk. These are just a few examples of the usefulness of recombinant DNA technology.

On the other hand, the same technology can pose dangers as well through unanticipated effects of a transgene, such as the herbicide-resistant soybean becoming a weed that would be hard to kill or by passing on the resistance to a related plant, which could then become a pest. Some people fear the misuse of the technology to create extralethal biological weapons or produce a super race of humans. Others are more concerned with the social and ethical dilemmas stemming from the use of genetic information available through recombinant DNA technology. An all too real scenario less fraught with technical details than transgenic weeds is the anticipated threat to personal privacy and possible genetic discrimination resulting from increased genetic testing for legitimate but unrelated reasons.

Genetic engineering employs the processes used in living cells to reprogram the machinery of other living cells. It seeks to redirect the chemistry in some or all of the many cells making up an organism to change it in some particular way. The changes, usually one or a small number, are made at the level of the information stored by the cell that tells the cell what to do and when by altering the long strands of DNA code known as the genome. The Human Genome Project was completed several years ahead of schedule. The scale of this task can be envisioned by imagining the DNA in the genome stretched end to end extending around the equator of the earth. One chromosome would then be 1,000 miles long and one gene would be one-twentieth of a mile, roughly the length of a football field. Within that gene could lurk a change of one nucleic acid base for another, a mutation, in the span of one-twentieth of an inch! Knowing the sequence of the genome is roughly equivalent to having a telephone book with the names and addresses of all of the molecules of the body as well as a map showing how they are connected. Such information could potentially be misused or have unforeseen unpleasant consequences. The practical and ethical implications of the availability and use of human genomic information is part of the genome project. Guidelines for its use and for the protection of individuals and discussion of the various issues are available from the genome project (National Human Genome Research Institute). The potential for misuse is already receiving the attention of numerous watchdog genetic resource support groups as well as federal and state governments (Council for Responsible Genetics).

Biomanufacturing

A host of medically important products once only laboriously extracted from animal tissues or human cadavers is now produced by genetically engineered microorganisms or animal cells. Prominent examples include insulin, erythropoietin, interferons, growth hormone, blood-clotting factors, antigens for vaccine production, and white blood cell proliferation inducers. Anticancer, antigraft rejection, and antiviral antibodies for treatments not requiring a host immune response are used in large quantities. Besides being free of potential human pathogens, many of these proteins are extremely rare. By using the human form of the protein, immune reactions to the proteins are reduced. This is one application of genetic engineering that has been an undisputed success.

Transgenic plants can be made to produce some of the parts of a vaccine, usually proteins made only by invading organisms. These can be extracted and used without the disease organism ever being present. Ideally, vaccination by mouth, as the Sabin oral polio vaccine is given now, although it doesn't always work for all vaccines, could be achieved by expressing the vaccine components in foods that can be eaten raw. Vaccines produced in food plants that didn't have to be extracted to be effective would be ideal in countries lacking access to medicines or refrigerated storage. Potatoes, alfalfa sprouts, cowpeas, bananas, rice, and tomatoes have been altered to produce the antigens for hepatitis B, measles, yellow fever, diphtheria, polio, cholera, and traveler's diarrhea by researchers at a number of universities and at Mycogen Corporation (San Diego, California). Similarly, the antibody molecules recognizing a vaccine can themselves be made in plants like other proteins to provide short-lived immediate protection. Other examples include an antibody against a bacteria causing tooth decay expressed in tobacco plants, extracted, and mixed with toothpaste to fight cavities, while a different antibody made in soybeans is being used to target drugs against cancers.

Agriculture

Humans have experimented with improving food production and food quality since they first domesticated animals and planted crops. Microorganisms were harnessed to brew alcoholic beverages for home and religious use and to process milk into cheese for unrefrigerated pantries. Selective breeding of plants and animals increased food supplies for a burgeoning population. With the advent of modern fertilizers, herbicides, and pesticides and improved methods of land use and livestock management, the Green Revolution of the 1960s hoped to feed the world.

Genetic engineering has entered into plant and animal breeding projects because of the speed with which desired changes in traits can be made, condensing generations of breeding and crossbreeding over years and generations into a single transfer of genetic information. There are limitations, however; in plants at present, such transfers are restricted to the manipulation of single or small numbers of linked genes (close together on a plant chromosome) and thus fairly simple characteristics. Complex multigenic traits involving large numbers of interacting genes are still out of reach and require traditional crossbreeding. Some changes can be made by transgenesis that will not occur through traditional breeding. For example, the species barrier to transfer of traits can be circumvented by moving the nitrogen fixation capability from legumes to corn or wheat to reduce fertilizer use. A host of desirable qualities including resistance to plant diseases, herbicides, various pests, salt and toxic heavy metals, freezing, drought, and flooding are being considered as targets for genetic engineering of plants. Protease inhibitor or starch-utilization inhibitor overexpressing plants show insect resistance. A cold-regulation gene switch has been identified that controls plant cell defense responses to low temperatures (Pennisi et al. 1998). Controlling cold response to increase plant tolerance could extend growing seasons and expand the regions supporting agriculture for some crops. Production can also be increased by improving the efficiency of photosynthesis, implanting nitrogen fixation or colonizing factors to attract nitrogen fixing bacteria, or controlling ripening of fruit. The nutritional content of major food and forage crops can also be improved.

The current generation of crop transgenes provides either herbicide tolerance (mainly glyphosate, bromoxynil) to reduce cultivation and spraying to remove weeds or insect resistance (mostly *Bacillus thurengensis* toxin, Bt) to reduce spraying of pesticides. The widespread adoption of this technology is for the major crops in the United States, which include corn, soybean, canola, cotton, and potatoes. The number of acres of herbicide-tolerant crops planted in the United States increased 20-fold between 1995 and 2000 to 100 million acres (Kalaitzandonakes 2003). The second generation of transgenic crops will likely address resistance to stress (drought, salt) as well as enhanced photosynthetic and nitrogen-fixing activity. More emphasis will be placed on industrial products, enhancing the nutritional qualities of foods, and production of medical products.

The rapid growth of genetically engineered agribusiness has ignited the smoldering resistance to the industrialization of the food supply by providing a fresh set of concerns specific to genetic engineering and the prospect of what some groups have called "Frankenfood." In addition, the business model being followed by the multinational agribiotech industry has stirred up resentment among nonfactory farmers and poor developing nations whose fragile economies fear corporate control. These social concerns exist alongside significant ecological issues, some of which are transgene-related and others of which recapitulate the worry during the Green Revolution over the ecological stability of large monocultures of genetically identical plants.

Industry

Biotechnology has played a role in industrial production of fermented products—cheese, yogurt, alcoholic beverages, and soy sauce, to name just a few—for centuries. Most of the world's supply of organic chemicals was produced by microorganisms before 1920 when Standard Oil of New Jersey began chemically synthesizing isopropyl alcohol (rubbing alcohol) from propylene, a petroleum product. Some fifty years later, the oil price rises of the 1970s and now the 2000s, along with the advent of genetic engineering, made biologically derived raw materials economically competitive. The controlled breeding of crop plants and animals in pursuit of high yields of particular industrial chemical building blocks such as plant oils with desired characteristics has continued unabated. Genetic engineering has allowed for quantum leaps in production efficiency of single products in a very short time compared with the years of crossbreeding required by traditional genetics.

Feedstocks

Transgenic plants have been the choice in many cases over microbes for new sources of feedstocks for two main reasons. Plants directly access the prime energy source, the sun, and large-scale cultivation and processing of plants is already practiced. Genetic engineering has provided a new wrinkle in the ability to increase yields of starting materials and to make available new classes of industrial building blocks. Genes that code for the production of materials previously obtained from petroleum or less efficient plant or animal sources can be inserted into either microorganisms or higher plants to enrich for industrially valuable products. The requirements for industrially useful materials differ significantly from the same materials in food products, especially economically, because they face intense competition from petro-derived products. Useful bioproducts, other than food, may include various oils and fatty acids containing from 8-22 carbon backbone chain atoms that are in demand for use in soaps, detergents, cosmetics, lubricant grease, coatings, plasticizers, drying oils, thermoplastics, and varnishes. Erucic acid produced in rapeseed plants engineered by Calgene is used as a lubricant and as a starting material for making nylon 13-13. Formerly, synthetic polymers used in containers and as textile fibers such as the biodegradable Biopol (polyhydroxybutyrate-polyhydroxyvalerate copolymer) are produced in a number of bacterial fermentation systems, recovering them from the harvested organisms. Greenpeace uses this copolymer instead of synthetic polyvinyl chloride for the plastic credit cards that they sponsor. Agracetus (Monsanto) has created "wash-and-wear" copolymers with cellulose by transferring the bacterial genes for the polyesters to cotton plants. Other components of the cell wall materials of plants such as the aromatic polymers comprising lignins and the sugar polymers forming the various types of cellulose and starch are also useful. They are building blocks for traditional chemical processes, fuels, or fermentation substrates for microorganisms to produce useful products. The obstacle to routine use of bioprocessing and bioproduction of industrial materials is making them cost-competitive with petroleum-based starting materials.

Bioconversion

Using specific genes or even multicomponent metabolic pathways from esoteric organisms cloned into "workhorse" strains of bacteria (*E. coli*), algae, or fungi (yeast) or higher plants (tobacco), normal cellular metabolic products can be converted into scarce drugs or precursor chemicals for industrial use. Molecules useful as medicines are naturally produced by specially adapted organisms in minute amounts under particular conditions. Metabolic pathways can be optimized and reconfigured to yield the desired product even if it was only a minor metabolite in the original organism. This is a rapidly expanding facet of genetic engineering that is predominant in the pharmaceutical production of medicines. Traditional chemical industry is also beginning to apply genetic engineering technology in place of more expensive chemical synthesis, notably of stereoisomers of chemicals, which enzymes can make in pure form.

Renewable Fuels

The production of biogas (50%–80% methane) from fermenting garbage, animal and human sanitary waste, and agricultural waste materials has been practiced on a small scale worldwide. A broad range of organic products can be converted by microbial activity through a very low-tech process. The main difficulty comes in expanding the process cheaply to industrial scale. The 5.6 billion m^3 of methane, equivalent to 5 million tons of petroleum, produced by Getty Synthetic Fuel is only a drop in the U.S. energy bucket. To achieve economic sufficiency in fuel production microorganisms are being engineered to use low-cost energy sources and to function at elevated temperatures in the presence of high concentrations of metabolites and at low oxygen concentrations.

Bioproduction of the ultimate clean fuel, hydrogen, which yields only water on burning, can be carried out by the green alga *Chlamydomonas* or the blue-green alga *Anabaena cylindrica* with energy from light (Benemann 1996). A two-stage process using photosynthetic bacteria is being tested at an Osaka power plant. Fermentation of organic wastes by bacteria in the dark can yield hydrogen mixed with methane as an adjunct to bioremediation to give a clean burning fuel. With all these technologies economics will drive the utility of the process. Present yields of hydrogen are in the 10% to 20% of input with economic sustainability hovering in the 60% to 80% range. Closer control of fermentation conditions and genetic and metabolic engineering of pathways will be needed to attain the breakeven point. Germany and Japan have invested significantly in this technology, whereas the United States lags significantly, spending about $1 million annually on research.

Biopulping

Chemical treatment of wood fibers is used to prepare wood pulp to make paper and to whiten it, an expensive, and environmentally damaging, process. A fungus that grows on wood, *Trichoderma virida,* accomplishes much of the same chemistry, partially breaking down the cellulose into a starting material for paper, producing as a by-product a sugar mixture that can be used to feed microbes to do other jobs. Again the problem lies in the industrial scale-up, speeding up the process and reducing costs.

Genetic engineering of the metabolic pathways in woody plants that produce the lignin fibrils and crosslink them together to give wood its strength and resilience has been the subject of much research (Baucher et al. 2003). Transgenic tree lignin pathways have been altered to fine-tune the properties of the wood fibrils to make them more suitable for pulping and to reduce drastically the use of treatments with chemicals or heating for which disposal is environmentally damaging. The effect of the changes in the physical characteristics of the wood for the ability of the modified trees to withstand wind, water, and insect pests is unknown. Environmentalist opposition to releasing modified trees into the field and fear of vandalism have thus far largely prevented trials designed to determine the consequences of the modifications.

Biomining

The recovery of copper and silver from mine leachings in the early Roman Empire was reported during the first century A.D. The natural action of indigenous bioleaching bacteria such as the common *Thiobacillus ferrooxidans* and *Leptospirillum ferrooxidans* solubilizes metals from their ores where they are generally present as oxides or sulfides. These organisms, which live in the ore deposits, grow best in highly acid solutions, pH 1.5–2.5 (neutral water has a pH of 7), using energy from either oxidizing sulfur compounds with oxygen or in its absence oxidizing ferrous (Fe^{+2}) iron to ferric (Fe^{+3}) iron. Carbon and nitrogen to make cellular substances come from atmospheric carbon dioxide (CO_2) and nitrogen (N_2), and phosphorus is obtained from soil mineral phosphate. Similar bacterial action can remove sulfur from coal deposits by conversion to sulfuric acid to make both a commercially valuable acid and fuel coal while reducing sulfate air pollution. Other bacteria such as *Thiobacillus thiooxidans, T. acidophilus,* and *Acidophilium cryptum,* which recycle some of the *T. ferrooxidans* products, often grow in association with *T. ferrooxidans,* making ore decomposition more efficient. Bacteria of the *Sulfobus* genus attack ores that are resistant to *Thiobacillus* action and thrive at temperatures approaching 80°C; these can work on ore deposits in situ, without excavating the ore. These organisms replace the high temperatures and high pressures of industrial processing plants.

Biomining can process lower-grade ores than is normally commercially feasible, although the process is slower. Controlled bioleaching can recover metals from low-grade ores or mine wastes (<0.5 % metal content) as well as more enriched sources; at the same time, it can minimize environmental pollution from naturally occurring leaching. It is already a commercially viable operation; in the United States alone, copper worth $350 million and uranium worth $20 million were recovered by microbial processes in 1985. Worldwide, 25% of copper worth $1 billion is biomined. Many metals, primarily copper and uranium, but also including cobalt, zinc, nickel, gold, and lead, can be obtained in this way. An annual worldwide value of $90 billion worth of biorecovered metals was projected by 2000 (Gorham International, Inc.).

Bioceramics

A characteristic of biological systems is their ability to organize physical structures. Besides their cellular materials, they can also organize the deposition of inorganic constituents such as bone ($Ca(OH)PO_4$-hydroxyapatite) and the remarkably intricate yet strong and resilient exoskeletons of shellfish ($CaCO_3$) as well as those of diatoms and sponges (SiO_2) (Drum and Gordon 2003). The structural details of these mineral shells have dimensions significantly smaller than components on current generations of microelectronic chips. This size range is known as nanodimensional, the realm of molecular-sized machines. Some grasses, such as *Brachiaria humidicola,* contain highly structured "silica bodies," which could be used for constructing submicrometer-sized structures. Control of microscopic structure like this would be a boon for industrial uses of silicon and ceramics as lightweight, inexpensive replacements for metallic parts in high-temperature or corrosive applications. They can also serve as templates for nanoelectronic components and sensors and as gratings for separating different wavelengths of light. Medical uses of bone substitutes require specific forms of the mineral to be biocompatible in reconstructive therapies. Biological control of ceramic microstructure is an emerging field (Mann and Ozin 1996).

Bioremediation

Overview of Air, Water, and Soil Cleanup

Plants have a remarkable ability to influence their immediate environment both directly and by the commensal communities of microorganisms that they attract and cultivate. Root systems of different plants permeate soil anywhere from a few inches (bluegrass), 1 to 4 feet (ryegrasses, fescue, clovers, and vetches), or 10 to 15 feet for prairie grasses and some wildflowers. Trees (poplar, willow, and eucalyptus) can extend their root systems even deeper. The vast surface area of the root system (rhizosphere) is home to the many kinds of bacteria and fungi that exchange carbohydrates from the plant for nutrients that the microorganisms liberate from soil components. At least 10 to 100 times as many microorganisms live in the rhizosphere as in dry soil without plants. Root systems introduce oxygen into the soil, which greatly speeds the metabolism of foreign chemicals in the soil by the microorganisms. Plants also secrete large quantities of carbon-based material into the soil that sometimes constitutes up to

50% of the material produced by photosynthesis, which precipitates many toxic materials such as heavy metals (mercury, chromium, copper, iron, lead, and zinc). They bind to the soil around the roots and concentrate there to protect the plant.

Contaminants spread through the leaching caused by diffusion of groundwater through the soil. Uptake of water by deep-rooted plants can prevent toxic plume spread by depressing the water table to below the contaminants. A 5- to 10-year-old tree can take up from 20 to 40 gallons of groundwater a day. Wetland systems employing ponds and lagoons with cattail, bulrush, and reed beds have been used for several decades to treat municipal and industrial wastes as well as agricultural runoff. A variety of federal environmental laws govern the use of phytotechnology for remediation. These include the Resource Conservation and Recovery Act and the Comprehensive Environmental Response, Compensation, and Liability Act. Along with this broad-ranging legislation the Brownfield initiatives, the Clean Water Act, and the National Pollutant Discharge Elimination permit system seek to regulate and validate the efficacy of bioremediation schemes. A variety of state and local statutes can also apply.

Along with their evolutionary tendency to fill every conceivable ecological niche, microorganisms have developed the capacity to utilize many sources of energy and nutrients in order to live and reproduce. This includes many materials that are toxic to animals and most plants, such as a wide variety of petroleum products and heavy metals such as mercury, lead, and uranium. They can also adapt to degrade a wide variety of human manufactured chemicals such as dioxin that are rarely produced and fluorocarbons that are not produced by nature. Using biological organisms to remove toxins and wastes from contaminated materials is called bioremediation. These living creatures can process trace contaminants present at the part-per-million level or lower in water, soil, and air, or destroy millions of gallons of spilled oil. Where do the organisms come from, and where do they go? What does genetic engineering have to offer such a talented resource?

Organic Chemicals

Organisms capable of metabolizing almost any organic chemical are already present in small numbers in the soil and water, a product of the constant struggle for nutrients and space. Living in intimate contact with their surroundings, they have evolved the means to deal with toxic materials in their neighborhood. With large numbers of organisms and heavy selection pressure by toxins, microbes that can't deal with the insult don't grow or die out, leaving those that can detoxify it to multiply and take their place. Microbes found near natural petroleum deposits are enriched in species that can utilize the hydrocarbons and aromatic chemicals that were created over millions of years by the decay and metamorphosis of prehistoric plant remains, some caused by microorganisms. They treat an oil spill as a bonanza and multiply rapidly, using it as a food source that the other bacteria in the soil cannot. Each type of organism can usually only effectively metabolize two or three compounds, so a diverse native bacterial population is an advantage over inoculation of spills with pure oil-eating cultures. In the Alaskan *Exxon Valdez* oil spill of 1989, an effective cleanup measure was to spray oil-soaked beaches with nitrogen- and phosphorus-rich fertilizer to encourage the growth of the indigenous oil-eating bacteria, which converted much of the oil to CO_2 (Swannell, Lee, and Mc-Donagh 1996). Although effective, bioremediation is considerably slower than intensive physical cleaning, depending on the environmental conditions. When the food oil runs out, the bacteria die down to trace levels again. Open water spills have not responded as well to bioremediation treatment. Cleanup of some 20 square miles of oil-saturated sand from sabotaged oil wells in Kuwait that threatened both water desalinization plants and the coastal life after the 1991 Persian Gulf War enlisted microbial help from the Kuwait Institute for Scientific Research near Kuwait City.

Commercial use is being made of microbes selected from natural environments. Two kinds of bacteria found in the soil of a paint landfill and a junk pile combined with a fungus are used at Kelly Air Force Base in San Antonio, Texas, to strip paint from airplanes without using solvents. Another bacterium degrades the paint.

Chlorinated aromatic chemicals are a considerable environmental liability in that they are carcinogenic, break down slowly in the environment, and tend to concentrate in fatty tissue and cause reproductive problems in birds and mammals. Organisms are being genetically engineered with combinations of metabolic pathways that break down such compounds more extensively and efficiently than the parental organisms in the soil.

Water-soluble Contaminants

Nitrogen- and sulfur-containing aromatic compounds generated by mining, coal tar and oil shale processing, wood preserving, and pesticide and chemical manufacturing processes endanger water supplies. These water-soluble molecules are not retarded by adsorption to soil components and pass rapidly into aquifers that feed human water supplies. Compounds of this type are substrates for a number of species of soil bacteria (*Pseudomonas, Corynebacterium, Brevibacterium, Bacillus, Nocardia*) that convert them to simple organic acids, ammonia, and sulfate ion, which then harmlessly enter the normal food chain.

Heavy Metals

Removal of heavy metal contaminants from soil and water is a different kind of problem. Because these materials are toxic in their elemental forms, they cannot be broken down into innocuous fragments but must instead be concentrated either for reclamation or for proper disposal. Microorganisms play a part in the natural release of toxic metals from ores and mine waste. Plants, in particular, the *Brassicaceae* (mustardweed) family, accumulate metals such as nickel, cobalt, copper, zinc, selenium, and lead in the above-ground part of the plant, in some cases up to 1% of its dry weight. Further improvements could be obtained by using genetic engineering to incorporate bacterial enzymes to reduce the metal ions to the uncharged, less soluble, and less toxic elemental metal. Highly efficient metal transporter molecules used by marine phytoplankton to recover trace metals from ocean water could be inserted genetically into land plants to increase metal uptake. Peptides binding toxic metal ions with high affinity could also be engineered into plants to increase their capacity.

Medicine and Health

Human Genome Project

As noted earlier, the nucleotide sequence of the human genome was completed ahead of schedule thanks to technological improvements in sequencing methods and in computer programs to overlap and align each of the thousands of small stretches of nucleotide sequence as they were determined. The project was a race mainly between the U.S. government–funded Genome Sequencing effort and the private company Human Genome Sciences, headed by Craig Venter (Shreeve 2004). An international sequencing project was also undertaken. The U.S. competitors announced the completion of the sequencing in a joint press conference at the White House in June 2000. The two groups used different strategies for organizing the sequence that they determined, which provided a check on the accuracy of the sequence and ironed out discrepancies.

The DNA sequenced in the genome projects was from a pool of DNA samples from a number of anonymous individuals, male and female. The European sequencing project used DNA from a pool of 270 individuals of African, Japanese, Han Chinese, and European descent. The U.S. efforts used a pool from individuals of European, African, American (North, Central, South), and Asian ancestry. Subsequent studies of multiple individuals suggest that less than 0.1% of the total nucleotide sequence differs among individuals, which would result in some 3 million sequence differences. Significant differences in sequence were not expected within the coding region of genes, the part that specifies the sequence of proteins and RNAs because that is controlled by selection pressure from the environment. The parts of the genome that vary most between individuals and ethnic groups occur in the vast regions between genes or sometimes in the control regions of genes. These variations, or polymorphisms, are largely responsible for subtle associations of ethnicity and other traits with certain phenotypes.

Genetic Testing

Genetic testing involves the search for specific sequences of nucleotides in DNA from individuals to see whether they harbor changes in that sequence that studies have connected with a disease or an increased risk of developing a particular disease. A number of technologies are available to do this. PCR amplification (mentioned earlier) of short target regions of chromosomes containing the areas of concern or binding of specific sequence probes to those regions has replaced the earlier methods employing restriction endonuclease DNA fragmentation patterns of those regions. Multiple genes or regions (or both) of the same gene can be tested simultaneously.

Although the technology for detecting changes in the genetic material has and will continue to advance, the question remains as to whether such screening should be done, particularly for diseases for which there are no treatments. Only in certain cases with genetically dominant mutations such as Huntington's disease, or where the individual has inherited two mutant copies of a key gene is the effect of the genetic change certain to be expressed. This ethical question may not have a blanket answer. For diseases affecting large enough numbers of people, studies may be able to correlate particular sequence changes in a gene with relative risk of developing the disease and possibly with age of onset. This would be useful for planning healthcare but would have implications for employability and insurance of the individual. The confidentiality of such information under control of the individual would then be critical to prevent discrimination.

Gene Therapy

Gene therapy is a medical treatment altering the genetic potential of cells. There are currently two types of possible genetic alterations. In one form, changes are introduced into nonreproductive (somatic) cells, often only certain types of cells. These changes affect only treated individuals and cannot be passed on to their offspring. The medical practice of gene therapy is in its infancy. Specific delivery of normal genes to the diseased part of the body, replacement of damaged genes, and turning on and off specific genes are still beyond reach of the "gene doctors."

Many people support the alteration of genes that cause a debilitating illness or deadly disease. On the other hand, the use of gene therapy to "adjust" prenatally other embryo characteristics has engendered much controversy. How many parents would opt for enhancing "good" attributes such as intelligence and physical features if such treatments were available? Gene therapy is also expensive, which would mean that only wealthy or well-insured people would effectively have access to the enhancements.

In the second form, germ line gene therapy, the permanent alteration of cells, including the reproductive (germ) cells—sperm and ova, causes the altered trait to be passed on to subsequent generations. Government guidelines and consensus in the medical community currently forbid these germ line changes. Besides formidable technical barriers to making permanent genetic changes in reproductive cells, people disagree about whether it is right to make permanent changes. Some people believe that while a patient may give consent to his or her own treatment, they question whether that individual has the right to choose for all succeeding generations who might have to bear the consequences of the alteration. This is countered by those who point out that essentially the same type of choice is made for future generations when a reproductive partner is selected. There is also the question of societal reaction to someone who had undergone germ line treatment somewhere in his or her pedigree. They worry whether such individuals would be stigmatized like someone who has undergone treatment for mental illness. Senator Thomas Eagleton, George McGovern's vice presidential running mate in the 1972 U.S. presidential election, was forced to withdraw from the contest when it was revealed that he had undergone mental testing. Conversely, others believe that correcting a deadly or debilitating disease condition to allow a person and his/her offspring to live a normal life far outweighs the risks and that not providing relief when it is available is wrong. Religious objections to changing "God's handiwork" are also a consideration for some people.

Pharmacogenomics

The application of knowledge of the human genome sequence to understanding variation in responses of individuals to pharmacological agents (pharmacogenomics) has been slow to gain acceptance. Part of this is a lack of information on large populations about genes that would affect response or metabolism. A major reason for this is that for economic reasons pharmaceutical companies are loath to subdivide their markets. Their emphasis is to encompass as large a patient population with a therapeutic agent as possible rather than restrict those would receive treatment. Only if a large enough fraction of the patients were predicted to be responsive to the therapy would this become attractive. Pharmacogenomic segmentation of patient populations, however, is attractive for running clinical trials of new drugs. Reducing the number of nonresponders in a study group decreases the total number needed to demonstrate a given level of clinical efficacy and thus the cost to run the trial. An example of this would be to select only subjects who express particular forms of drug-metabolizing enzymes that degrade, inactivate, or modify a drug so that it is excreted. This would eliminate efficient metabolizers of a particular drug, thus increasing the chance that the drug levels required to see a clinical effect would be attained. The trade-off is that with this type of trial, approval from the Food and Drug Administration might only be for demonstrated non-metabolizers. However, a small trial could give encouragement to a company to run a larger, more inclusive trial and perhaps to monitor drug levels to adjust dosages accordingly.

DNA and the Law

The use of DNA "fingerprinting" as evidence in court was thrust firmly into the public limelight during the O. J. Simpson trial in 1995. The legal system now accepts that when properly calculated, the probability of matching crime scene DNA to that from a blood sample can be estimated. Such conclusions bear in mind the somewhat different prevalences of variants of genetic markers in different racial and ethnic populations. Forensic DNA analysis was first introduced into evidence in 1986. Acceptance was not immediate, coming only after statistical arguments based on population genetics probabilities were established as valid, by defining the probability that two DNA genetic marker patterns appear identical simply by chance. The Frye Test, a 1923 U.S. Court of Appeals ruling for the District of Columbia Circuit, was used to judge when evidence supplied through recombinant DNA technology met the criteria for judicial technical evidence. It also indicated the general acceptance of DNA evidence in the relevant scientific community. Fingerprint and other forms of scientific analysis now commonly accepted had been required to meet similar standards. DNA evidence can be and has been excluded in some cases if the evidence was shown to have been collected or analyzed improperly, just as evidence of any other type would be treated.

DNA sequences differ between individuals at the level of approximately 0.1%, or around 3 million nucleotides out of the 3 billion nucleotides in the human genome. These differences are not scattered randomly throughout the genome. There are 13 DNA regions in the genome that vary from person to person. The probability of two persons having the same profile in all 13 regions is 1 in 1 billion. By comparison, the probability of eyewitnesses being able to identify a person correctly out of a lineup as having been at the scene of a crime is approximately 1 out of 2, or half the time. Fingerprint analysis, the previous gold standard for identification of individuals, uses a series of features called Galton points after a 19th-century founder of biometrics, Francis Galton. The number of matches required depends on the jurisdiction and the country. The Federal Bureau of Investigation

(FBI) currently does not set a minimum number of matches constituting a positive identification.

The current FBI standard is 13 short tandem repeat sequences that are determined and then stored in a database. CODIS (Combined DNA Index System) is a software program that operates local, state, and national databases of DNA profiles from convicted offenders, unsolved crime scene evidence, and missing persons authorized by the DNA Identification Act of 1994. All 50 states participate in CODIS at some level. For older samples or tissue that lacks cell nuclei such as bone, hair, and teeth, the DNA sequences from mitochondria, a subcellular organelle found in all cells, are analyzed. This DNA is inherited from the mother and is compared with the mitochondrial DNA sequences from a relative on the individual's mother's side of the family.

Biological Warfare and Bioterrorism

Biowarfare

In 1969, President Richard Nixon signed a decree unilaterally renouncing biological weapons, purging them from the U.S. arsenal. The Second Asilomar Conference on Recombinant DNA in 1975 was very concerned with the possibility of genetic engineering of biological weapons of mass destruction. They concluded that "construction of genetically altered organisms for any military purpose should be expressly prohibited by international treaty." In 1975, most countries in the world signed just such a treaty banning production, possession, and stockpiling of biological weapons and toxins. Oddly, the treaty did not apply to chemical weapons (McDermott 1987). Although genetic engineering would appear to be ideally suited for creation of devastating disease weapons, many scientists doubt the possibility of intentionally developing a genetically engineered Doomsday Microbe or an Andromeda Strain virus. Lack of understanding about the causes and limits of disease makes it difficult to deploy and to control the effects. Among developed nations at least, the military finds such weapons to be too difficult to make effective and to limit their action to an opponent to be useful. Since 1975, most nations of the world have renounced the production, storage, or use of biological weapons.

Inexpensive to produce and easy to hide, biological weapons have been called the "poor man's atomic bomb." Nevertheless, use of such weapons by militarily weak nations and suicidal releases by reckless individuals who disregard the consequences remain a possibility. There was concern during the Iraq conflict in 1991 following the invasion of Kuwait that in desperation Iraq would use biological and chemical weapons. The U.S. Department of Defense has reportedly maintained a research effort to provide defense against biological weapons that might be used against the United States. Periodically, there are questions about whether the research is straying into the arena of offensive uses. The issue is so politically charged that the military is careful about even appearing to be involved in genetic engineering. This policy has been marginally successful because of secrecy surrounding the work, which breeds distrust. Public suspicion is fueled by periodic revelations. In 1992, the Yeltsin government admitted that the Soviet Union had maintained a "standby" program in biological weapons in the 1970s. An apparent leak of four strains of deadly anthrax from Military Compound No. 19 in April 1979 into the city of Sverdlovsk, Siberia, was the first proven escape of an infectious agent from a military biological laboratory resulting in fatalities. Recombinant DNA technology ultimately provided the proof through PCR analysis of frozen autopsy tissue from victims (Hoffman, 1998). Biological warfare has remained a highly emotional issue particularly in Iraq, and concern over its possible use was given as one reason for the U.S.-led invasion of Iraq in March 2003 looking for "weapons of mass destruction" under the peace treaty ending the 1991 Gulf War.

Bioterrorism

The United States received a taste of the potential impact of bioterrorism in the form of a white fluffy powder that spilled in October 2001 from mail addressed to members of Congress. These spores of the bacterium *Bacillus anthracis,* a bacterium found in soil in low numbers, germinated to infect 7 people with a skin form of anthrax and 11 with the more dangerous inhalation respiratory disease. Although this particular strain of the anthrax bacillus was sensitive to the antibiotic Cipro, the disease, which is uncommon in developed countries, was recognized too late for some victims. Five of the inhalation anthrax victims died, having come in contact with mail that had been processed along with the envelopes laced with the spores. Although the number of people actually affected by the spores was small, the fear raised and the speed and ease of spread was a sobering demonstration of the impact of a bioterrorist attack.

The U.S. mail system was disrupted for months as post offices were tested for contamination, and the Congressional Post Office was closed for more than a year for decontamination.

Subsequent studies showed that the spores had been partially weaponized. The suspension had been prepared in such a way as to be stable, easily spread, and fine enough to penetrate normal lung defenses. Although the bacterial strain was eventually found to be similar to that used in several university and military microbiological laboratories to study the disease process, the perpetrator(s) were never discovered. The sophistication of the material initially implicated foreign biowarfare programs, but domestic sources are also suspect. A number of researchers' careers were destroyed as accusations flew, and the low level of control of potential biological agents that might already be in the hands of terrorists was uncovered. Steps were taken before and after the establishment of the Office of Homeland Security in 2002 to control access to potential biological agents. Scientists found their research on a number of disease-causing organisms suddenly exposed to intense scrutiny and were burdened with bureaucratic record keeping and layers of security. Some of their work was sharply restricted or banned altogether. None of this will ensure that biowarfare agents will not fall into the hands of potential terrorists.

On the other side of the coin, a new biotechnology industry sprang up to respond to new government initiatives funding the development of new methods of detection and decontamination of biological and chemical agents. Discovery of new antibiotics potent against favorite bioweapon organisms and treatment protocols is being encouraged. Community plans and training to detect and limit the spread of biological agents are being implemented. The impossibility of total protection of air, water, and food resources from terrorist acts leaves restricting the impact as the best option.

The specter of genetic engineering in the service of terror is a very real threat. The technology for legitimate applications is widespread, although its application to potential bioweapon agents is more complex. Small laboratories are sufficient for development work, although production is more specialized. Saddam Hussein was widely suspected of having a bioweapons program, although clear evidence was hard to come by in the wake of the U.S. invasion of Iraq in March 2003. Resistance to current antibiotics and the combination of toxic agents in a single organism are only a few of the possible twists that could increase the impact of an agent. Because the purpose of terror is not necessarily to decimate an opponent but to disrupt society by introducing fear and uncertainty, biological weapons are a natural choice. They are invisible, tasteless, odorless, self-propagating, easily delivered, and hard to detect, and only small amounts of the substances are required. Genetic engineering also has the potential to combat bioterrorism. Engineered organisms can produce new forms of antibiotics, signal the presence of toxic chemicals, or interfere with the ability of bioagents to infect the host. It can be used to construct new vaccines against biological agents that would be effective against bacteria or even against viruses, which are not affected by antibiotics. Humans, livestock, and agricultural production could potentially be protected, although the effectiveness against an unknown agent would be uncertain.

Computers and microprocessors are a glittering testimony to the accomplishments of human technological development. Creating something that never existed before by processes not seen in nature was a monumental achievement. By contrast, genetic engineering used processes employed by natural systems for billions of years to promulgate life and adapt to changing surroundings, harnessing their power to modify existing biological systems. In doing so, genetic engineering reached a threshold that requires decisions; just because a technology *can* do certain things does not necessarily mean it *should* be used in those ways. The details of the technology of genetic engineering are confusing to many, particularly to those who have little understanding of biology, which automatically generates suspicion and fear. Combined with questions the new technology asks of deeply held principles of ethical and moral conduct, this powerful mix occasions a great deal of discussion among technologists, business interests, medical scientists, theologians, ethicists, politicians, and common citizens. Governments have the responsibility to manage this conflict to advance all interests and to protect the rights of the parties while using the parts of the new technology that will solve human problems. As one might imagine, this is not an easy task.

Conclusion

New technologies generally give rise to opportunities as well as to questions and concerns. Genetic engineering is no exception. There are clearly contributions to improving the human situation with enhanced medicines and possibly new treatments in store as the Human Genome Project nears maturity and the data become available to the research and pharmaceutical sectors. Less clear are the gains from bioengineered agriculture, which promise industrial feedstocks from renewable plant sources and high-quality foodstuffs using fewer chemical fertilizers or herbicides and pesticides and leaner, fast-growing animal products. To achieve this, ecosystems will be exposed to genetically modified organisms on a massive scale well beyond any testing. Will they be able to coexist, and can an ecosystem recover from infusion of these types of foreign species? Even for desperately needed bioremediation to clean up our environment, the same ecological risks apply.

Beyond questions of safety and the environment rise myriad social issues that cannot be settled by experiment and observation. Genetic engineering carries a considerable economic impact as well. Countries able to support the necessary research and development will have an advantage over those that do not, and they can economically pressure the situation, creating dependency and further separation of the haves and have-nots.

Maintaining the privacy of an individual's genetic information will be a paramount issue in the face of potential use of that information by employers, insurers, and others. The growth of DNA databases will make controlling accessibility more challenging. Tests for genetic diseases will continue to proliferate, and genetic counseling will become increasingly important as more treatment options become available, including gene therapies.

In a sense, Aldous Huxley's *Brave New World* is already here. The difference is that we have the opportunity and indeed the obligation to see that it is done right. Huxley's work of fiction warns us of what we must be aware and what we should preserve. Through careful evaluation of the effects of our newfound technology on the environment and on what we believe the essence of humanness to be, we need to make generally acceptable choices. This will require both education and compromise because a number of the issues poise at the border between rationality and faith. Such differences are hard to resolve directly. They are, nonetheless, solvable at some level though probably not entirely to everyone's satisfaction, and this will ensure a continuing and, in the end, a healthy debate.

Now that the background of the development of genetic engineering has been laid out and the technologies briefly explained, it is important to consider the ramifications of their introduction.

References

Baucher, M., C. Halpin, M. Petit-Conil, and W. Boerjan. 2003. "Lignin: Genetic Engineering and Impact on Pulping." *Critical Reviews in Biochemistry and Molecular Biology* 38: 305–350.

Benemann, J. 1996. "Hydrogen Biotechnology: Progress and Prospects." *Nature Biotechnology* 14: 1101–1103.

Council for Responsible Genetics. http://www.gene-watch.org/.

Drum, R. W., and R. Gordon. 2003. "*Star Trek* Replicators and Diatom Nanotechnology." *Trends in Biotechnology* 21(8): 325–328.

Friend, S. H., and R. B. Stoughton. 2002. "The Magic of Microarrays." *Scientific American* (February): 44–53.

Hoffman, I. "Piece Found in Anthrax Mystery. Lab Sheds Light on Russian Leak." *Albuquerque Journal,* February 1998.

Kalaitzandonakes, N. 2003. *Economic and Environmental Impacts of Agbiotech: A Global Perspective.* New York: Kluwer/Plenum.

Mann, S., and G. A. Ozin. 1996. "Synthesis of Inorganic Materials with Complex Form." *Nature* 382: 313–318.

McDermott, J. 1987. *The Killing Winds: The Menace of Biological Warfare.* New York: Arbor House.

National Human Genome Research Institute. http://www.nhgri.nih.gov/Goeff Spencer.

Pennisi, E. 1998 "Transferred Gene Helps Plants Weather Cold Snaps." *Science* 280: 36.

Shreeve, J. 2004. *The Genome War: How Craig Venter Tried to Capture the Code of Life and Save the World.* New York: Knopf.

Swannell, R. P. J., K. Lee, and M. McDonagh. 1996. "Field Evaluations of Marine Oil Spill Bioremediation." *Microbiological Reviews* 60(2): 342–365.

Velander, W. H., H. Lubon, and W. N. Drohan. 1997. "Transgenic Livestock as Drug Factories." *Scientific American* (January): 70–74.

2

Problems, Controversies, and Solutions

Since the Second World War and the great destruction wrought by that armed conflict, there has been a rising movement to consider more carefully the consequences of new technology. The atomic bomb's vastly increased destructive capacity over previous weapons and the critical role of scientists in producing that weapon and subsequent generations of ever more powerful devices jolted the public and many scientists to reconsider the social responsibilities of developing technology.

Problems and Controversy

Within this framework, significant technological advancements and advantages have to be weighed against the moral, ethical, environmental, economic, and safety impact. In the United States, some of these areas are traditionally considered the purview of government regulators and policy makers, whereas others reflect personal choice of values or religious faith. Still others require interaction and dialog between nations and groups of nations. Genetic engineering applications potentially span all of these areas, tapping into some of the deepest convictions. Disagreements over certain issues are rooted in systems of religious faith. This makes it particularly difficult to have rational discussion and dialog. Hot-button issues of this type in the United States are human reproductive cloning and human embryonic tissue uses, which include embryonic stem cell research.

47

Proper government oversight and legislative networks can, in theory, address many of the concerns in areas of health, privacy, environmental and economic impact, and forensics. Following through on these responsibilities, however, is not simple and requires thorough investigation and much discussion to arrive at agreement on the proper ratio of cost to benefit for society and for individuals.

Safety of Genetic Engineering

Many of the early genetic engineering scientists worried about the escape of modified organisms into the environment, their effects on people, and the possibility of upsetting the global ecology. Their apprehensions were shared by a significant number of nonscientists and public officials who were concerned for public health and envisioned creation of new uncontrollable forms of life. A general lack of understanding of what could and couldn't be done with genetic engineering was the cause of much unrest. Federal guidelines first issued in 1976 and updated at intervals have regulated the types of genetic engineering experiments. Although public debate has largely subsided over the years of apparently safe application of the technology, watchdog groups such as Jeremy Rifkin's Foundation on Economic Trends continue to monitor and challenge the release of genetically engineered organisms into the environment and marketing of food products from genetically engineered organisms through the Pure Foods Campaign (Rifkin 1998).

Genetic Engineering and Health

The unique contribution of genetic engineering technology to medical science is the ability to alter specific portions of the genetic code of an organism to repair genetic defects in an individual to cure genetic diseases. Until now people had to accept the genetic background they inherited. In theory at least, this may no longer be the case. Related DNA technology makes testing for mutations rapid and accurate. Benefits of detection and correction of aberrant DNA will come to fruition only after much complex experimental and clinical investigation. Recombinant DNA technology also provides tools for understanding better how the body functions in health and disease and for discovering new treatments. This process is explained in a *Scientific American* article (Haseltine 1997). Cellular and animal models that mimic human and animal diseases can be engineered for the testing of new medicines.

Genetic engineering, like other technologies such as microelectronics, has the potential to revolutionize the way things are done for the betterment of society, but there are costs to be borne as well. Along with an improved standard of living for many, microelectronics ushered in automation that gradually marginalized unskilled labor with serious economic and societal consequences. Significant changes of a different kind in society would also be expected with recombinant DNA technology. Some people assert that difficulties arising from the new genetic technology can be anticipated and controlled. Others fear that dangers of unrecognized proportions loom for genetic engineering. Potential scenarios range from disruption of the earth's ecosphere with consequent crop failures and widespread starvation, to the development of devastating "germ warfare" weapons. Less biologically devastating but potentially more socially disruptive would be the use of genetic information to assess health risks resulting in discrimination against people in obtaining insurance coverage or employment on the basis of their genetic heritage. These are distinct possibilities unless controlled by legislation. Ensuring protection of the privacy of an individual's DNA and the ethical implications of choosing the genes of future generations with some gene therapies are emerging as significant concerns.

Genetic Testing

Genetic engineering technology can be used to probe the heredity of individuals. That genetic knowledge carries both health-related and societal consequences. Genetic testing is normally performed on three groups of individuals, the first of which is unborn children (prenatal testing) to determine whether they have inherited a fatal or severely debilitating disease. Adults who want to know whether they inherited a particular genetic trait for a disease that occurs late in life or whether they could pass a genetic disease on to their offspring are a second population. Lastly, asymptomatic children are tested in a few cases for certain later-developing diseases, but such testing is generally discouraged until the age of majority when the young adult can decide for herself or himself whether to be tested.

Genetic analysis for the presence or absence of altered genes only confirms that the altered gene is present in an individual. Except for a relatively small number of disorders in which the gene product is dominant, inevitably expressing the disease although the time of onset is uncertain, only a risk is associated with the presence of the altered gene. For nondominant, or recessive, genes, to what extent and when, if ever, the disease will occur are unpredictable. For multigenic traits such as heart disease in which more than one gene is involved the individual genes are often considered risk factors rather than direct causes. Many environmental modifications such as introducing a diet low in cholesterol and saturated fat can intervene to prevent disease from developing by reducing the risk.

Prenatal testing for biochemical or genetic defects by analyzing amniotic fetal cells, placental tissue, or fetal blood is routinely used (300,000 cases/yr in the United States) where the family history or advanced maternal age (35–40 years) indicates an increased risk of fetal abnormality such as Down syndrome. Every baby born in a U.S. hospital is presently screened for at least two anomalies, phenylketonuria and congenital hypothyroidism. Most are tested for a larger number, depending on the state. Title XXVI of the Children's Health Act of 2000 undertakes to improve and strengthen newborn metabolic screening in the United States. At present, biochemical testing for the expressed effect of an altered gene, rather than the gene itself, is preferred. This is because determining risk due to the presence of an altered gene is harder to interpret because not all gene defects are fully expressed. Finding the biochemical hallmarks of the disease before it causes significant damage is a sure way of knowing early. A negative biochemical test could change later depending on the underlying genetics, but the effects are more readily noticed once the person is older.

Eventually, prenatal genetic tests may be used for the same disorders in place of the biochemical tests because they are easier to perform and are more sensitive. Interpreting the results will likely remain uncertain. Genetic counseling accompanying the testing provides advice to the parents on the possible outcomes of the pregnancy and to inform them of their alternatives. Many of the genetic diseases screened for are presently untreatable, leaving either termination of the pregnancy or raising a short-lived or handicapped child with heavy burdens on the family and the child. Some people question whether the genetic counseling about risk is understood as a probability or a certainty by the parents. Although the counselor is only trying to make information available and not give advice, it is difficult to ensure complete neutrality. Other people believe that counselors *should* provide guidance. They reason that if society becomes responsible for the afflicted individual, then society should set some standards for quality of life.

After a person is born, what are the implications of a diagnosis of an incurable disease of unpredictable severity? How would that alter a person's life? Huntington's disease (HD), a neurodegenerative disorder, typically becomes apparent in adulthood after the disease has been passed to the next generation. In the absence of a treatment for the disease or accurate prognostication of severity, HD family members often do not want to be tested, suffering the insecurity, and hoping that they will experience only a mild form of the disease. HD parents would have a hard time coping with the knowledge that they have passed on a serious genetic disease to their child.

Health Privacy

Particularly in the United States, personal privacy is highly prized and zealously guarded. There will be a greater need for confidentiality protection as more potential disease genes are identified. Although medical records are traditionally considered confidential, is a person's genetic code any more confidential than fingerprints? Use of anonymous genetic markers for personal identification could be considered a similar but more conclusive method as long as the markers don't eventually become related to identifiable personal characteristics or to disease loci. At that point, the information would constitute a medical record, which then would dictate a different level of confidentiality. Ostensibly, information from a soldier's DNA sample can only be used for identifying remains. Police can obtain DNA samples under warrant only to exclude suspects in a crime, but what guarantees that these will be the only uses of those samples? Few states have moved to bar unauthorized access to DNA databases.

The advances in genetic technologies triggered a debate about how privacy of individuals should be protected. Is it a question of protecting civil liberties, which require one standard of safeguards, or is genetic information a form of health information, which requires a different treatment under the law? Because genetic information will always be linked to health, it is regulated under the umbrella policy issue of health information. The Health Insurance Portability and Accountability Act of 1996 does many things besides what the name implies. It compelled the health insurance industry to make certain changes to provide more patient control of their medical information by April 2003, or April 2004 for small health plans. These rules are in addition to restrictions by individual states. Unlike the federal law, the states tend to treat genetic information differently from other medical information and use a variety of means to achieve more patient control to safeguard the genetic information.

Insurance and Employment

The insurance industry stays in business by spreading out the risk of paying settlements over many policyholders, some with a higher risk for payment than others. Health and life insurance companies contend that genetic testing simply adds more certainty to the standard medical testing and family history commonly used in actuarial analyses to determine that risk. They argue that this knowledge would allow more focused assessment and lower premiums for individuals. However, this policy would result in other individuals, who from their genetic background and no fault of their own, are at a higher risk for requiring expensive healthcare or early death facing unaffordable premiums. Many would likely be refused coverage altogether. By arguing that genetics are beyond the control of the individual, lawsuits have forced insurance companies to cover individuals with genetic predisposition to a disease and the payment of medical expenses incurred by a child born with a diagnosed genetic disorder. In response, insurance companies and employers may be taking a closer look at premiums for people whose *chosen* lifestyle exacerbates their risk.

Insurance cost- and productivity-conscious employers increasingly attempt to use genetic screening to select their workforce to cost them least to insure and miss the least amount of work due to illness. Fair employment practices as set forth in Title VII of the Civil Rights Act of 1964 preventing a variety of discriminations are the bedrock of genetics protections. The American Disabilities Act of 1990 (ADA), which forbids discrimination based on disabilities that do not affect performance, is important for people with expressed genetic diseases. Genetic predisposition has been considered protected against discrimination by the Equal Opportunity Commission since 1995, which invoked the ADA, but the U.S. Supreme Court has been less sympathetic to this interpretation in a number of cases. In 1992, Wisconsin became the first state to forbid discrimination in employment or insurance coverage based on an individual's genetic readout. A 1996 federal law already bars insurers of group plans from considering genetic predisposition a "preexisting condition" in order to limit or charge more for coverage, unless the condition is clinically evident. The statute does not address self-insured individual health plans, which account for 10% of health insurance. Laws in thirty states restrict the use of genetic information in the workplace.

Forty-eight states restrict employment-related medical exams and set the confidentiality standard but stop short of forbidding access to medical information, including genetic information. An employer can seek certain genetic information related to the ability of an individual to perform job-related functions or to workplace safety.

Sixty percent of life insurance, forty percent of disability insurance, and nearly all long-term care insurance policies are individually underwritten and thus not covered under the federal protections. In 1997, spurred by the cloning through nuclear transfer of several mammalian species, the federal government was considering a wave of antigenetic discrimination and genetic privacy legislation. By early 1998, nearly 250 bills in 44 state legislatures proposed limits on access to genetic information. Forty-five states restrict what health insurers can do with genetic information, but only seventeen states address the use of genetic information for decisions on life, disability, and long-term care insurance. As in the federal case, the proposed regulations would only govern group, not individual, policies. In formulating legislation, regulators need to balance safeguards for individuals with excessive adverse effects on the insurance market.

State laws tend to restrict rather than ban the use of genetic information for certain kinds of coverage other than basic health insurance. These include life, long-term disability, and long-term care, possibly because they are seen as less of a necessity. However, the underwriters usually must show actuarial data that people who had the particular genetic condition required significantly more payments than would be expected from the overall population. Genetics and long-term care insurance will become an increasingly visible issue as the U.S. population ages and as more genetic risk factors are discovered for protracted diseases of the elderly, such as Alzheimer's disease.

Human Cloning

Although the debate raged over the safety of recombinant DNA technology at the Asilomar Conference in 1973, the issue of human genetic engineering was purposely omitted from discussion. Because it was so far from the realm of possibility at the time, and realizing that strong opinions were held on both sides of the issue, the organizers, probably wisely, chose to keep this volatile topic from derailing the conference. Better to concentrate on issues that could be addressed and some conclusions reached, although the implications of the new DNA technology for human engineering remained in the back of everyone's mind. The serious dialog that society needed to have was simply postponed for 24 years. Controversy reared its head in the form of a sheep created from an ovum, an egg cell, by replacing its nucleus with a nucleus from a cell with its genome committed to be an udder cell expressing only udder cell genes. Although the procedure used is actually nuclear transfer, the media immediately christened the accomplishment "cloning." Presumably, this was for its added shock value, because that word is not mentioned in the original published scientific article describing the procedure (Wilmut et al. 1997).

The deprogramming of a differentiated genome had been accomplished with nonmammals in a number of laboratories, but mammals just hadn't worked. Previously believed to be irretrievably committed as an udder cell, the transplanted nucleus, after it had been treated in a prescribed manner to initiate the process, reversed the udder cell programming and became embryonic. It was then capable of dividing to produce cells of all the types needed to form a lamb when implanted in a surrogate mother. Although the success rate was low, 1 out of 277 tries, the seemingly impossible had been achieved, and soon other nuclear transfers with mammals including two rhesus monkeys were reported, many of which were under way at the same time as the sheep. Thus, it is likely that this was a case of the technology maturing to the point where it is now *technically* feasible to do a similar procedure with human ova and nuclei. There are unsubstantiated reports from individuals and organizations that human clones have been produced.

Although not requiring the recombinant DNA technology that is the subject of this handbook, it is easy to foresee (or imagine) further developments that would allow certain "repairs" or "enhancements" to be performed on the donor nucleus before transfer, making this truly human genetic engineering. The nuclear transfer technique was perfected to make available animal clones of superior livestock (already practiced by artificially inseminated embryo implantation) and to facilitate production of transgenic human proteins in these animals. However, nuclear transfer cannot compete economically with the embryo transfer method for livestock. Embryo transfer implants embryos produced by artificial insemination into surrogate mothers. Stem cell technology, which like nuclear transfer "cloning" does not itself involve alteration of genetic material, could also become a vehicle to engineer changes in characteristics, although current uses involving cell or tissue regeneration do not create a new individual.

The fervor of the public response, as well as the flurry of ethical condemnation and legal activity, was reminiscent of that greeting the advent of recombinant DNA. A moratorium on nuclear transfer experiments was immediately issued in the United States and a raft of federal and state legislation proposed and position statements from scientific societies and religious groups banning human cloning issued. The Food and Drug Administration (FDA) stepped up to fill a perceived regulatory vacuum. Any human cloning requires FDA approval. Internationally, nineteen members of the Council of Europe signed the European Convention on Human Rights and Biomedicine in January 1998, part of which bans "any intervention seeking to create human beings genetically identical to another human being, whether living or dead." Germany, which already forbids any human embryo research, and Britain, which was involved in the sheep cloning and aims to continue with its tradition of protecting the freedom of scientific inquiry, did not sign the protocol.

Religious Implications of Genetic Engineering

Implicit in much of the discussion in this chapter is the concern that the application of genetic technologies be fair in respecting the privacy of the individual and that whatever burdens are imposed fall reasonably equitably on various groups in the population. There is a question in some people's minds as to whether some or even any of the technology *ought* to be applied at all. The general uneasiness felt by some people over technology's ability to interfere with the order of the natural world was stimulated by the somatic cloning of sheep and other animals. The generally held abhorrence of analogous "cloning" experiments with humans has cut across a wide range of societies, religions, and philosophies. Much the same sort of agreement surrounded the passing of the United Nations Declaration on Human Rights in 1948 when, according to David Tracy in an essay in *Clones and Clones: Facts and Fantasies about Human Cloning* (Nussbaum and Sustein, 1998), "Jews and Christians, Muslims, Buddhists, Hindus, Taoists, Confucianists, and people of several indigenous religions found it possible to agree with the practical list of human rights—but each for their own ethical, metaphysical, or religious reasons." In the case of cloning, it would be difficult for these divergent parties to agree on the details of the concepts. For now, they agree on prohibiting one type of manipulation. The concordance may not extend to other procedures. A case in point is the use of animal tissues, genetically modified to improve cross-species tolerance or not, in transplants. Pig and baboon hearts have been transplanted into humans whose own hearts have failed and for whom a human donor was unavailable. The Old Testament (Leviticus 19:19) forbids the crossbreeding of distinct plant or animal kinds, prohibits sexual relations between humans and animals, and proscribes eating an animal before the blood had been drained from it. Some Christians, as well as Jews and Muslims, argue that this type of statement strictly condemns biological intermingling of humans and animals. On the other hand, because genetic material is not exchanged in xenografting—interspecies transplantation—and because many Christians interpret that human beings have been reconciled to God through the death and resurrection of Jesus Christ, thus superseding the ceremonial regulations of the Old Testament, many feel that these medical procedures are allowable. Some argue that even the observed barrier in nature to crossbreeding between species—indeed, the definition of species—is a human construct and not a "natural" dividing line. Over the millions of years of evolutionary time, these barriers constantly change and therefore appear to exist only for a given time, particularly in related branches of the evolutionary tree, such as mammals that share much of their DNA sequences.

Still, others would point out that the order of biological systems has developed, whether by self-organizing principles or through the divine intervention of a purposeful Creator, into a smoothly operating whole that should not be disturbed by "mere" human beings because we cannot know the full extent of the consequences of our interventions. Cross-species transmission and adaptation of diseases is a very real possibility, especially worrisome for infectious agents that we might not know about yet. This comfortable, conservative principle of not rocking the boat has appeared in several contexts in considerations of the applications of genetic engineering.

The Judeo-Christian ethic uniquely values human life, and killing of humans is forbidden except in punishment or in time of war; it places strictures on appropriate animal use for human benefit such as for food. Do recipients of xenografts, say, pigskin replacement of burned tissue, become any less of a person? Because grafting does not carry on into the next generation, concerns over succeeding generations and how they will be viewed by their contemporaries should not arise. This differs from germ line gene therapy, which will modify all generations and is currently forbidden in medical practice. Although many people accept the breeding and raising of animals for human food, some people have difficulty accepting such practices for organs or tissues, considering this yet another assault on the rights and feelings of animals.

The cloning of humans also raises religious and moral objections about the dignity of a person—treating every human as an individual. This is a consequence of the widely held belief that no human being should be treated as an instrument to an end. Cloning an incurably ill child as a "replacement" or creating a genetic double to serve as a tissue donor, or even to gratify the ego of someone who wants to create a clone of themselves, violates this premise. Illogically, having a child "the old-fashioned way" for the same purposes seems to be less of an issue. Such apparent inconsistencies riddle discussions on ethics and morality in general, but despite them some sort of consensus needs to be reached. A practical view suggests that ethicists and the public need to concentrate on what to do with cloning rather than trying simply to ban it. Many people feel that if it is possible, it will probably be done somewhere eventually, and making sure that agreed-on standards are present to govern the application offers the best means of excising meaningful control. Past experience banning other societal behaviors (e.g., prohibition) has taught us that much.

Embryonic Stem Cells

The controversy over embryonic stem cell research is a result of the rekindling of the abortion debate. Because derivation of embryonic stem cells using today's technology by definition involves destruction of the embryo, it is subject to all of the bioethical objections of those opposed to abortion. This includes excess embryos created for in vitro fertilization, of which some 400,000 are thought to exist in the freezers of in vitro fertilization clinics worldwide (Lamb 2004). Only 20 of the original 78 embryonic stem cell lines existing before the Bush administration deadline of August 9, 2001, are useable to scientists who say they need many more lines to learn how to direct their growth. Because the technology is available and can be applied anywhere in the world, the most productive recourse would be to provide the proper ethical oversight rather than to try to ban its use.

Legislation aimed at restricting the application of embryonic stem cell technology currently regulates research paid for by federal funds. Private individual or company practice is controlled by the states, the jurisdictions of which differ. Numerous bills and resolutions in both houses of Congress have been discussed, further defining and supporting research under the guidelines laid down in 2001 (S. 723/H.R.2059).

There are many highly visible lobbies promoting stem cell technology, including actor Michael J. Fox who suffers from parkinsonism and former first lady Nancy Reagan, whose husband, former president Ronald Reagan, died from Alzheimer's disease. Both of these neurodegenerative diseases could potentially benefit from successful embryonic stem cell therapies. In November 2004 California voters supported a $3 billion bond referendum to support embryonic stem cell research. Venture capitalists are investing significantly in new firms that are targeting embryonic stem cell therapies. It is too early to evaluate the utility of embryonic stem cell technology.

Impact of Genetic Engineering on Health and Privacy

Biological science, particularly when applied to medicine and pharmaceuticals, has been utterly transformed by the new genetic technology. Genetic engineering has been tremendously en-

abling in experimental science and in the discovery of new medicines over and above the scarce hormones, growth factors, and
antibodies that can be produced on an industrial scale by direct
application of the technology.

On the other end of the spectrum is the decided lack of
progress on gene therapy. Gene therapy is something that recombinant DNA technology has the potential to do that other treatments cannot: remove the root cause of a disease by replacing the
deficient gene that caused pathology. Initial concern over human
engineering was defused by a general agreement to ban germ line
gene therapy for the present. The debate was reawakened in February 1997 with the generation of a sheep cloned by nuclear transfer from an adult differentiated cell. Although not using recombinant DNA technology per se, the feat served to reopen the
discussion. Despite the approval and implementation of hundreds
of somatic gene therapy clinical protocols, convincing evidence of
a significant clinically useful effect remains to be shown. Seasoned
practitioners point out that this is exactly what would be expected,
especially because these treatments were therapy of last resort for
most patients. Clinical medicine has always been a slow, trial-and-
error process. Changing the pathophysiology was not going to be
as simple as popping a new gene into the diseased tissue. Proponents of gene therapy would now agree that they are still learning
how to get genes into the right place at the right time for the proper
length of time and that the process is much more complicated than
was originally anticipated.

Although no worldwide ecological cycles stand in peril from
human genetic testing, the developed nations are experiencing
considerable upheaval and public debate over its social implications. Those countries without the infrastructure or scientific resources for their own recombinant DNA programs may initially
be spared the trauma of adjustment to a wealth of personal genetic information, but they will eventually have to face up to its
implications as filtered through their own societal norms and sen-
sibilities. They will have the opportunity to observe the debate
and types of solutions attempted by the nations that are facing the
dilemmas now and will be able to fashion their own versions. The
technology or personal or public safety in this case is not at issue,
but the use of the personal information obtained by individuals,
the government, and the business and industrial sector to determine personal opportunities and relationships is at stake. There is
no technological fix, no straightforward way to ensure that people

are treated fairly while retaining the advantages of having genetic information available. It is a social problem with greatest impact at the moment on the developed nations that are trying to deal with it. Linda Bullard articulated the dilemma:

> Like nuclear power, genetic engineering is not a neutral technology. It is by its very nature too powerful for our present state of social and scientific development, no matter whose hands are controlling it. Just as we would say, especially after Chernobyl, that a nuclear power plant is just as dangerous in a socialist nation as it is in a capitalist one, so I would say the same thing for genetic engineering. It is *inherently* Eugenic in that it always requires someone to decide what is a good and what is a bad gene. (Bullard 1987)

Much controversy centers on the privacy of a person's genetic information and how that privacy will be maintained in an environment where such information, which is out of the individual's control, could be used in hiring and firing decisions and in determining insurability risks. Ensuring control over access to information stored in databases has been an issue since the advent of the widespread use of computers to store and access that information. Although overall security has been reasonable—so far—there are more than a few public instances in which sensitive information has been obtained despite the safeguards. With the projected creation of extensive databases and DNA banks, such as DNA databases of convicted criminals and military personnel for rapid identification, better control measures able to handle larger amounts of data stored in more places become imperative. At the same time, just such repositories of genetic information, with personal identifying characteristics removed, are indispensable for research into understanding diseases and fundamental biological processes in the 21st century.

Foreseeable consequences of the ability to detect genetic disorders are questions about how such information should be used as well as to whom the testing should be available. Whether to test in the first place, it is generally agreed, depends on whether there is a high rate of conversion to the clinical disease (i.e., Huntington's disease, Down syndrome, and cystic fibrosis), which could be used as a base for reproductive decisions. Testing would also be indicated if an effective treatment for the disorder is available. The consequence of the inappropriate use of genetic testing results is the a priori limitation of personal opportunity on the basis of projected genetic potential. This last issue particularly concerns people who have been diagnosed with a genetic disorder but who do not yet show clinical signs of impairment. It also affects their families and caregivers. The possibility of disastrous confusion of genetic probability with certainty for most genetic diseases makes a difficult and unsettling situation for the individual with a positive diagnosis even worse. A genetic predisposition to a disorder before symptoms appear is not the same thing as a clinical disorder where the person or the physician notices the changes due to the disease. The complex interdependence of factors causes many genetic disorders to develop only partially if at all into clinical disease. Nevertheless, the afflicted individual is constantly under the pressure of trying to live a reasonably normal life, all the while looking for the first signs of the disease. Should they tell a potential spouse of their condition and should that person also be tested to avoid a union that might produce children with an aggravated genetic risk? Parents may be tempted to divert educational and other family resources to "healthy" offspring with their presumed better chance to make better use of the opportunity.

Mammalian cloning—that is, nuclear transfer as it is currently practiced—does not make use of the recombinant DNA techniques that mark the genetic engineering discussed in this volume. However, the consequences are similar in some respects and as such deserve some comment here. The term *genetic engineering* originally referred to human engineering, which would include reproducing an organism from a single cell rather than its DNA complement. The organizers of the Asilomar conferences in 1973 and 1975 agreed to exclude the topic of human engineering because of the technical infeasibility of such experiments at the time. They also recognized that the extremely emotionally charged nature of such an issue would quickly dominate the public mind, and no consensus would be possible on the more mundane but more pressing issue of the safety of recombinant DNA technology matters. The issues of human cloning using somatic cells are closely allied to those of human germ cell gene therapy, which has by consensus of the medical community been banned. Besides any theological sensitivities that might be offended, in either of these cases a person would be created who could, at a future time through no action of his/her own, be subject to discrimination or ostracism depending on the course of societal values.

Possibilities for Action

When it comes to human health and the research that leads to better and deeper understanding of human health and disease, there is near consensus that genetic engineering has contributed immensely to furthering knowledge and practical pharmacological treatment. One needs to look no further than the enormous private investment in biotechnology companies and in the corporate acquisition of biotechnology capability. Most of this investment activity is in the pharmaceutical and diagnostic sector. People believe that the technology will produce saleable products. On the down side, the miracle gene therapies promised by the new technology have not as yet lived up to their potential in any meaningful way, outside of some judiciously interpreted proof of concept cases.

The arguably most profound impact of the genetic revolution may well be on the ethics of some of the applications of genetic engineering technology. In theory, it is possible to overcome the technical objections to safety, ecological disruption, and economic distortions from unequal access to the technology. The philosophical issues are yet another problem. Some of the ethical questions being posed now have been disputed vehemently among intelligent people for centuries, regardless of their religious persuasion, without resolution. Genetic testing raises the question of what (as well as who) is the individual and what rights does that person have to knowledge about himself or herself, and what rights do others have to that same information? What are the rights of future generations, and what responsibility does one bear toward descendants? The United States is moving toward legislative restriction of both access and use of genetic information for insurance or employment selection. Other countries feel differently about the relative importance of the individual and society and are less restrictive. Outside of this arena, in the United States genetic testing used to determine the potential of predisposition to major genetic disease is presently regulated by a series of medical guidelines depending on the age of the subject and on the prognosis for intervention in the disease. This is still a very murky area that will continue to expand as more markers for genetic disease are identified.

Nuclear transplantation, popularly dubbed "cloning," of humans and human stem cell replacements have leapt to the fore in the public notice as primary issues in genetic engineering, which is increasingly being eyed as human engineering. Not strictly genetic engineering in the sense of recombinant DNA technology, it is not difficult to imagine future "enhancements" that could be applied. Although not yet feasible for humans, the flurry of legislative activity all over the world designed to control or ban nuclear cloning and stem cell research attests to how close such activities approach what many people feel is the limit, the edge of their humanity. The challenge will be in setting the boundaries on exactly what is going too far without disrupting the beneficial parts of genetic engineering as applied to these new technologies. Finally reduced to fundamental issues of philosophy that are not amenable to proof and reason, this will be the burning question for this millennium.

Impact of Genetically Modified Foods

An issue is whether food products genetically engineered by recombinant methods should be required to be labeled as such and be processed and sold separately from the same products produced through traditional breeding programs. Public health concerns (side effects or allergies), religious proscription, and personal preference considerations supporting special labeling on one hand are arrayed against stigmatizing biotechnology for no demonstrable reason and boosting protectionist economic tactics by countries that lack a significant biotechnology infrastructure. Little useful information about actual product safety is provided to guide consumer decisions by such labeling of engineered crops.

Genetically modified foods, referred to as GMO (genetically modified organisms), are even a bigger issue in Europe than in the United States. In October 1999, the European Commission's Standing Committee for Food set a legal threshold of 1% genetically modified product contamination for imported agricultural products destined for human consumption. Because the same processing equipment is used to handle both GMO and regular products, this limit can be unintentionally exceeded. In the absence of data showing harm caused by GMO foods, U.S. processors (and the farmers who supply them) feel unfairly discriminated against. Genetically modified soybeans (Monsanto) and maize (Ciba) have been targeted by the European Union, bringing that organization into conflict with the World Trade Organization, which is responsible for overseeing the General Agreement on Tariffs and Trade. This agreement states that imports can be banned only on scientific grounds. Jeremy Rifkin's Foundation for Economic Trends, by way of the Pure Food Campaign, maintains a generally critical stance on genetic engineering. This group feels that labeling of all genetically modified produce should be required and that any resulting conflict with trade law should automatically be superseded by potential public health concerns that could be associated with the genetic alteration. Traditionally cross-bred crops, even those resulting in the same characteristics, would not be so regulated. This discrepancy is often pointed out by proponents of genetically engineered or enhanced foods who feel that such labeling needlessly stigmatizes their products, because harmful effects have not been demonstrated.

Genetic Engineering of Food Animals

Genetic engineering of food animals has been less well developed, restricted primarily to providing supplements of native growth hormone to cattle and pigs. Treatment of cattle with growth hormone protein produced by biotechnology increases milk production 10% per day, and animals reach market weight sooner with leaner meat. Although the hormone given is chemically identical to the natural hormone, there is much controversy over whether products from treated animals should be labeled differently from regular animals in case problems develop. The European Union still does not allow sale of products from animals treated with recombinant growth hormone.

In many areas of the world, the only source of high-quality protein is fish taken from the ocean, lakes, or rivers. Some scientists feel that increasing the amount of edible fish is not just convenient but necessary as commercial fish catches decline along with the natural fish populations due to overfishing. In fishery laboratories, sockeye salmon, catfish, trout, striped bass, flounder, tulipin, and other species are being given extra copies of fish growth hormone implanted in their DNA (transgenic fish). These superfish grow more than 10 times faster than regular fish, require less food per pound of body weight, and are adapted to be raised in high-density "fish farms." These farming advantages have raised concerns about what might happen to the ecology of natural systems if the faster-growing and reproducing transgenic fish escaped into the wild fish population.

Environmental Safety

There is considerable disagreement over whether genetically modified or transgenic plants grown on a massive agricultural scale represent the same or greater risk to the environment than those traditionally bred (Nottingham 2002). Few controlled studies on the actual risks have been carried out, and then only under a limited set of environmental conditions. Potential problems are seen for many of the proposed genetic modifications. Acquired resistance to engineered pesticides such as *Bacillus thurengensis* toxin (Bt) would negate the benefits of the transgene in crop plants as well as reduce the effectiveness of the Bt spores used by organic farmers for insect control. Cross-pollination and transfer of herbicide resistance (glyphosate, bromoxynil) to related weed plants could bring problems in weed control that wouldn't be outweighed by the use of small amounts of safer herbicides. Viral-resistant plants containing viral coat proteins (squash; Asgrow Seed Co.) have the potential to harbor virulent recombinant viruses with new properties, although the actual extent of pathogen evolution in large-scale transgenic crops is unknown. Other approaches or targets would not suffer this flaw. Antisense ribozyme against rice dwarf virus would avoid this problem. Genetically altered fruit ripening (enhanced ethylene production (DNA Plant Technology) or cell wall hydrolase antisense DNA (Calgene) would not be expected to have harmful effects on the environment.

Although comprehensive studies of a transgenic plant entering a wild population to become a weed or to disrupt ecological interactions among other plants have been lacking, prevented sometimes by the very activists who demand the results of such trials, lessons from history allow some estimates. Taking all known instances of introduction of nonnative species, a species becomes established in the new ecosystem 10% of the time. About 10% of these newly established species go on to become a "pest" species, so 1% of the time a newly introduced species becomes a pest. Although introduced species are more likely to be genetically different from the species around them, and thus less likely to interbreed, the population dynamics of the first generation of transgenic plant products (herbicide- and pest-resistant) do not differ markedly from their equivalent nontransgenic brethren. In a nonselective environment (no herbicides present or infestations of a monoculture), they would have no advantage and would remain at low levels. Crops such as transgenic soybeans, a nonnative species, and those that have no or few wild relatives are even less likely to be a problem. On the other hand, the spread of new genes from cultivated trees is more likely to be a problem because their pollen is transmitted over larger distances and genetic mechanisms in many tree species favor outbreeding.

Because of the ease of pollen dispersal, transgenic crop plants have been found in nontransgenic fields. Possible problems with *Bt*-toxin–laden pollen being eaten by monarch butterfly caterpillars have also come to public attention. Collateral effects like this are a real concern. Agribiotech companies haven't improved the image of the new technology by prosecuting farmers for license infringement when transgenic plants, likely cross-fertilized from windblown pollen, were found among their fields of nontransgenic plants in the vicinity of neighbors who had licensed the transgenic variety. Certified organic farmers and those who wish to sell standard nontransgenic crops are at risk of having their crops contaminated through no fault of their own. Some of these issues could be addressed if the transgenes were designed to be excluded from the readily dispersed pollen, although this is a more challenging approach. Perhaps this adjustment will be built into future generations of modifications.

Agribusiness Control of Agriculture

One of the differences between the Green Revolution and the Gene Revolution is that private plant-breeding firms have taken the lead in technology development in the latter. Government-sponsored programs now play a much smaller role in developing new plant varieties (Evenson 2002). Many are relegated to projects conserving the natural diversity of the germplasm resource and are poorly funded. Most research and development plant breeding, genetically engineered or traditional, is centered in the developed world where the genetic engineering technology is highly advanced and, importantly, where the intellectual property laws are clearest and most rigorously enforced. To protect costly investments, companies have for many years produced hybrid varieties of some plants (maize, sorghum) that require farmers to purchase new seeds each year for maximal yields.

Still, many farmers in the United States save seed from the previous year's crop to resow the following season. Up to 80% of farmers in developing countries use retained seed. New systems are in development that will further reduce the ability of farmers to reuse seed. They are called genetic use-restriction technologies (GURTs). So-called terminator strains are variety-based GURTs that will not grow from replanted seed. Trait-based GURTs will have properties of a standard variety but will require the purchase of a complementary product to get the enhanced production. These are high-tech solutions to what companies see as unauthorized use of their products.

Farmers cite the uncodified Plant Breeder Rights originally started in Europe and spread to the United States more than fifty years ago under which farmers could replant seed and to use freely any new varieties that might develop under their cultivation as long as they didn't sell seed (Swanson 2002). Farmers had traditionally been involved in the selection of varieties of crops under their care to obtain plants that were adapted to local conditions or displayed other superior characteristics. The Union pour la Protection des Obtentions Végétales (UPOV) codified limited Farmer's Rights for private replanting of registered seed in 1978. This protection was removed in 1991 as a result of objections by the biotechnology industry. The same agency in 1961 had provided exclusive marketing rights for registered plant varieties, similar to the patent system, which does not cover plant varieties. These restrictions are unevenly enforced despite the World Trade Organization's Trade-Related Intellectual Property Rights Agreement (1994) because many countries lack the infrastructure and funds to do so. Enforcement does not work because the users (farmers) are not in the same country as the plant breeders. Most agritechnology firms are multinational companies based primarily in Europe and in the United States.

A major conflict with a familiar dividing line is shaping up. The two sides are the rich, developed nations that hold the technology and the poor, developing nations in need of increased agricultural productivity to provide income for their farmers and food for their burgeoning populations. Some relief, although not a solution, may come from a generics industry expected to spring up in the next decade as patent protection lapses on the original technologies.

The ecological consequences of the widespread use of genetically modified organisms in agriculture and in industry remain largely unexplored. No application seems entirely risk-free. Even bioremediation, which is directed to redress environmental damage inflicted by industry and a consumer-driven society, is fraught with possible ecologic nightmares. Potential scenarios range from the rampant spread of pest and weed resistance, cold tolerance, massive crop failures and starvation, contamination of food crops with plants engineered to produce industrial chemicals, and destruction of the world's biodiversity and genetic reserves. Intermixing of genetically engineered foods and those derived from traditional breeding methods could expose consumers to unanticipated allergens or even potential toxins. These dire possibilities are not all trumpeted by a zealous and vocal few individuals with a dread of any technology or any change, although there are a number of such groups and individuals in the public eye. Various incarnations of these issues and evidence for a measurable effect can be found in governmental and research reports, although, of course, the true extent of their impact is unknown.

Those considering the positive aspects of genetic engineering point to maintaining and increasing food production for the burgeoning world population while decreasing the reliance on chemical herbicides and pesticides as laudable goals, worthy of a certain amount of risk. Opening previously unusable land with engineered drought-, salt-, cold-, and heat-resistant plant varieties could be augmented by plants with more efficient photosynthesis and enhanced nutritional content. Fast-growing, feed-efficient, lower-fat food animals with enhanced processing characteristics would increase the high-quality protein available.

Possibilities for Action

The agricultural impact of genetic engineering is likely to be only slightly less revolutionary than in medically related fields. Just as everyone gets sick, everyone needs to eat. Genetic technologies have the potential to increase food crop production further, using less fertilizer, fewer herbicides or pesticides, all in poorer soil with less and poorer-quality water during an extended growing season in harsher climates. Not established yet and greatly feared, though, are the potential ecological consequences of mass culturing of plants that could pass their favorable genetic properties to wild-type plants, which would then become weeds, disrupting both human applications and world ecology. The Environmental Protection Agency is responsible for regulating the release of transgenic organisms in the United States. There are also significant economic changes that would be incurred as well as social changes in the farming profession.

Particularly strident voices have been raised over marketing of transgenic crops, especially for human consumption. Vocal groups of consumers typified by the Pure Foods Coalition oppose the introduction of transgenic crops or products from plants or animals produced with the assistance of genetic engineering technology, citing potential health hazards from unnatural combinations of gene products. Resistance to transgenic foods and transgenic imports in Europe claims similar roots, but accusations of economic protectionism have also been made. At this point the perceived danger is a matter of personal preference, with little or no convincing data presented. Legislation on labeling foodstuffs known to be free of transgenic products would avoid stigmatizing biotechnology products for unproven dangers and would resemble similar labeling for organic products.

Ecological disaster caused by the escape of transgenic organisms is the most feared potential consequence of exposure of engineered organisms to the environment—and the hardest to provide evidence against. Adequate testing in the ecosystem in which the organism will be used is rarely done because of the expense and time involved. Understanding of large ecosystems is also poorly developed at present. Although no significant escape into the wild population has been observed for transgenic plants, this does not reassure some people. Meanwhile, enhanced risk-versus-benefit analysis is being carried out and releases are restricted until more is learned about their ecology.

Inextricably tangled with the "engineered" versus "natural" controversy is the widespread planting of genetically homogeneous plant populations, which expanded greatly with the Green Revolution. The potato blight in Ireland (1845), wheat "red rust" in the United States (1954), and the corn leaf blight in the United States (1970) are pointed examples of a population crash when a small number of varieties are being grown. In the 1980s, 70% of U.S. corn was in six cultivars while a single potato cultivar monopolized the Netherlands. This is a problem currently encountered with traditionally bred high-yield hybrids and is shared by genetic engineered varieties. Monocultures displace indigenous species and lower the diversity of the gene pool for those traits that are multigenic or not yet assigned to a particular gene. Such populations are particularly vulnerable should a pest or weed overcome the plant resistance and sweep through the crops. On the other hand, in developed nations farmers maximize crop yields and reduce cultivation requirements by planting the most productive varieties. It is unrealistic in a market-driven economy to expect otherwise. The danger of massive crop failures with consequent food shortages from a pest circumventing resistance or other methods of control is a constant threat. Some feel that the possibility is adequately being dealt with, but others believe that disaster is just over the horizon. In some sense, the dilemmas of today with genetic engineering recapitulate those presented by the Green Revolution in the 1960s. The desertification and aquifer depletion that have accompanied some high-intensity agriculture and continue today are seen by some as a prime example of the lack of proper management.

Although many people consider the use of genetic engineering in certain health-related situations as a necessity, its application in the production of food is much more controversial. At first glance, such an attitude is incongruous with the centuries of widespread plant and animal breeding to increase yield and resistance to disease. On the other hand, it is an example of human reticence to accept unfamiliar technology in everyday situations in which traditional solutions are available. The conception for many people is that there is nothing inherently better about the engineered food products and that the consumer and environment are put at unknown risk simply for the convenience and profitability of factory farm producers and manufacturers of processed foods. Particularly disturbing to some is that global economic conditions are changing such that choice among the options may not be freely available to all societies, particularly developing nations and the third world. The worldwide implications for large-scale application of plant and animal genetic engineering will be complex and will pit the needs and aspirations of the third world against the lessons learned by the industrialized nations as the developing nations struggle for their place in the sun. Suman Sahai, the convener of the Gene Campaign in New Delhi, India, summarized this perspective in *Genetic Engineering News* (May 15, 1997, issue): "There is little reason for people in food surplus countries to become excited about the biotechnology route to increase the yield of wheat or potato. But can we in India have the same perception? Is it more unethical to 'interfere in God's work' than to allow death from hunger when it can be prevented?"

Impact of Bioindustrial Engineering

The large-scale application of industrial bioengineering is likely to have an impact on the environment similar to that of genetically modified foods. In comparison to most medical applications of genetic engineering, industrial uses typically involve several orders of magnitude larger amounts of materials, sometimes running into the millions of tons. Working on such a scale while maintaining close confinement and monitoring of the stages of production with genetically engineered organisms is manageable but requires significant investment. Escape or accidental release into the environment from these larger populations could potentially disrupt the ecosystem. Using strains of organisms containing debilitating genes that put them at a disadvantage compared with organisms that are naturally in the environment provides a margin of safety if they escape culture containment, although it is often pointed out that such features are not foolproof. Similar concerns exist for both modified microorganisms grown in huge vats and for modified higher plants. The latter situation involves cultivation of thousands of acres of recombinant plants growing in fields next to plants that might crossbreed and transfer undesirable characteristics. The plants could also be infected with viruses or microorganisms that might exchange genetic material, and the plants are exposed to insects that might also disperse genetic material in unimagined ways. On the other hand, the products being expressed for industrial applications generally do not confer a selective advantage, unlike pest or weed resistance, so there is less weed potential for these systems

Situations such as biomining, in which microorganisms are released or enriched for use directly in the environment, clearly have considerable potential for causing ecological problems, although many of these microorganisms are already present and have adapted to living in the deposits, albeit in small numbers before manipulation.

Possibilities for Action

Industrial applications of genetically modified organisms include the extension of the traditional culture of microorganisms in closed systems such as sealed vats. Other systems are more controversial. Utilizing genetically modified plants to produce industrially useful building blocks such as certain oils or mixed ester fibers grown on a large scale runs into some of the same environmental problems of other large-scale transgenic plant propagation. Reducing the use of fossil fuels for industrial production in favor of a renewable resource fits with the trend toward ecologically responsible technologies and could be a significant contributor in the next century.

Biotechnology itself has become a substantial industry. The spirit of entrepreneurship still runs high in the United States in that most of these companies are small with almost 80% having fewer than 50 employees. Many were started at academic institutions and were often nurtured in university "incubators" by administrations hungry for new sources of income as federal higher-education funding dwindled. Although providing a pipeline to convert academic knowledge to practical application, the rush to obtain patent protection for ideas or products and the direct connection of academic scientists to industry has drastically affected the university-industry relationship and raised questions for many people about the supposed "disinterested" involvement of many academics and the role the government has played in funding the research. In Europe and the Far East, such connections between the universities and industry are more common than in the United States. In any event, the profound change in the university-industry relationship in the United States was an unpredicted outcome of the genetic engineering revolution.

Impact of Bioremediation

Destruction and cleanup of toxic materials on-site without creating massive dangerous waste storage areas using nature's own mechanisms—it sounds too good to be true. And so it is; by its very nature, bioremediation involves the large-scale release of a few types of organisms into the environment, possibly genetically modified microorganisms or plants. It can include controlling groundwater flow into or out of a contaminated area with plants that send roots to different depths in the soil. Control of releases and their effects on the local ecology such as exchange of genetic material with native populations or depletion of groundwater are issues that have and will continue to limit the dissemination of this technology. Regulatory agencies, local governments, and citizens will have to wrestle with the conflicting needs for detoxification of air, water, and soil, the need to protect fragile ecosystems, and public unease with genetic engineering technology. Many of the same questions that face other uses of biotechnology in the open environment are also germane here. Naturally occurring asphalt-eating bacteria are suspected to cause significant damage to roads, particularly where weather and heavy loads cause cracking, increasing microbial access. What is the possibility that super paint-eating or rubber-degrading strains could change to grow in a normal environment, attacking houses, cars, and machinery?

Possibilities for Action

Genetic engineering holds the potential for affecting a massive cleanup of the environment of the various toxic and noxious byproducts of industry and our high standard of living. Bioremediation of toxic waste in land, water, and air by either indigenous organisms or by engineered ones can be effective and relatively inexpensive without processing factories or moving large amounts of materials by treating them in situ. By their nature, however, these techniques require release of large numbers of organisms into the environment without being able to control their spread and disturbance of balanced populations of other organisms. They could even get into underground storage tanks or accelerate road destruction. Microbial ecology is even less well understood than terrestrial or aquatic ecology, so it would be even more difficult to predict release outcomes than in the agricultural case of a transgenic corn plant containing an herbicide resistance gene. Further study of microbial ecology could provide needed information to allow the design of biological containment systems that could provide for safe release.

Engineered higher plants with enhanced capabilities to deal with pollutants or attract microbial communities would be easier to contain than microbes and could provide much needed remediation capabilities. In contrast to many other uses of genetic engineering in which "ruthless corporate giants" and "greedy individuals" are perceived to be risking the public health and safety of others for their exclusive benefit, bioremediation has the potential to provide a positive impact for all parties.

Impact of Genetic Engineering on DNA and the Law

In a period of less than a half dozen years, the use of DNA information to identify individuals has moved from grudging acceptance to a highly sought link in the chain of evidence. It has been particularly useful in clearing suspects of charges, even if they had been previously convicted on the basis of other evidence. One result of this newfound acceptance is a large backlog in the processing of samples. Analysis is now more complete and the results more reliable than before with the introduction of a larger number of new markers. There are still issues over whether and how long DNA samples may be retained by law enforcement agencies and under what conditions they may obtain the DNA.

The nonforensic-related use of the DNA information deposited in a general database from a criminal investigation, however, has become a concern to a growing number of people. Some consider "fishing" for a suspect by randomly matching crime scene analyses of DNA markers to a general DNA database to be a search without a warrant, a violation of individuals' civil liberty. The FBI, as of January 2003, had DNA profiles on more than 1 million convicted violent offenders and 48,000 from crime scenes not connected with an offender. An early concern was that the markers used for typing in that DNA database might be linked at some later date to a genetic trait or behavior that then would make it illegally disclosed confidential medical history. With the human genome sequence completed, the markers chosen are not in gene coding regions and thus unlikely to be informative with respect to a trait or behavior.

Where Is Genetic Engineering Today?

How do the scientists and physicians, industry, universities, ecological activists, politicians, and finally the general public view the place of genetic engineering in the world today? What have been the gains and the losses over the nearly 30 years since the recombinant DNA revolution began? What are the hopes and fears for the future now that genetic technology is touching people's lives through its increasing use in providing food and medical treatment? The interconnection and interdependence of these sectors of society have increased greatly over the years. They are no longer isolated endeavors. Changes in one area have repercussions in all of the others.

Scientists and Physicians

Hopes were initially high for gene therapies to repair major genetic disorders such as muscular dystrophy and cystic fibrosis as well as to provide cures for cancer. Although numerous clinical protocols for cancer and genetic disorders have been approved and carried out, as of 2004, genetic therapies have not been hugely successful and appear to add little to conventional medicine. A variety of practical problems have stalled progress, although in principle the therapies show the intended effects. Much more development is needed before genetic treatments become part of the physician's armamentarium. The National Institutes of Health (NIH) Recombinant Advisory Committee, which is entrusted with approving gene therapy protocols, issued a report in December 1995 concluding that clinical efficacy had yet to be conclusively demonstrated for any gene therapy protocol despite anecdotal claims. Thus far, there has been consensus in the medical community not to engage in germ line genetic therapy in which the afflicted individual and all his or her progeny will carry the genetic change.

Genetic engineering has been successful in producing bioactive proteins with pharmacological activity, particularly with white blood cell hormones to stimulate the recolonization of immune cells after chemo- and radiotherapy in cancer patients and with human insulin. Although not the "magic bullets" that some had predicted, they are valuable and useful therapeutics.

Genetic engineering has completely revolutionized certain areas of biological scientific research, particularly in its application to medicine. Its impact goes far beyond the production of medical biologicals (insulin, interleukins, growth hormone, granulocyte stimulating factor, and monoclonal antibodies), which involve the relatively straightforward production of scarce biological proteins. The tools for today's biomedical research wholly embrace recombinant DNA technology. The data flowing from the Human Genome Project and other organism genome sequencing efforts are rapidly helping to define targets for new medicines and may eventually change the way medicine will be practiced. Pharmacogenetics will identify groups of patients that are genetically programmed to process medicines differently, which should reduce the incidence of side effects. Ironically, knowing more in greater detail than ever before about the human genome may lead to a more holistic manipulation of cellular and organismal properties to attain health.

Industry

The modern biotechnology industry has been the main beneficiary of genetic engineering. Biological technology has been practiced on an industrial scale since the middle 1800s, when it was primarily limited by production engineering capabilities. The introduction of molecular biology avoided this obstacle and placed the limits only on biological feasibility. Before the 1960s the term *genetic engineering* referred to human engineering, which at that time was science fiction. Today, *genetic engineering* and *biotechnology* are used nearly interchangeably, although biotechnology generally retains the broader definition.

The use of genetic engineering of plants and microorganisms to provide improved, cheaper feedstocks from nonpetroleum-based materials as well as specialty chemicals has been a natural outgrowth of the traditional biotechnology industry. Rises in crude oil prices make this method of production more economically attractive. Cultivation of large numbers of genetically modified plants raises some ecological concerns, but because the modified characteristics for industrial applications do not generally favor survival, there is less worry about escape into wild populations. Microbial production of useful chemicals including pharmaceuticals has continued over the years with little public concern. Conventional fermentation of genetically engineered microorganisms to produce some products is more readily confined and escape can be monitored, so there is less issue with this application.

Genetically engineered foods, on the other hand, have attracted a great deal of attention and will continue to be a lightning rod for activist concern and activity. Expansion of the cultivation of genetically engineered crops into underdeveloped countries raises questions about inadequate ecological testing and exploitation of the economies of these nations lacking significant capabilities in biotechnology. In this scenario, developed nations would provide the market but would keep underdeveloped nations dependent on the technology. Lack of intellectual property protection in many of the underdeveloped nations is cited as hindering the transfer of applied genetic engineering technology to those countries. The United States held off signing the United Nations–sponsored Treaty for the Protection of Biodiversity until 1993 until it won some concessions on intellectual property protection. These issues were addressed to some extent in the Trade Related Aspects of Intellectual Property Rights (TRIP) agreement of the World Trade Organization, which came into effect in 1995.

The genetic engineering industry likely will grow but at a more normal pace without the blockbuster expectations of the early years. In the near future, smaller biotechnology service-based companies will be established to support various forms of genetic testing and DNA forensics emanating from advances in medical understanding and as a result of the mountain of information provided by the Human Genome Project.

Universities

The development of genetic engineering has caused a realignment of the relationship of the research universities and industry. The many individuals and small enterprises involved in genetic engineering have blurred the distinction between inquiry to understand more about how things work (basic science), which is what normally is carried on by research universities, and applying that understanding to create a saleable product (applied science), normally done in industry. Federal funding by NIH, the largest source of research funds outside of defense, favors—in fact, expects projects to show—medical or health applications. Basic science discoveries are now patented and publicized, often before their soundness and utility have been established. Such behavior impedes the free flow of ideas and defeats peer scientific review of research results designed to maintain high scientific standards. It also interjects the question of conflict of interest between individuals trying to commercialize their findings and the research community. Meanwhile, many people feel that academic researchers whose work is paid for by public funding from the NIH, National Science Foundation, or other government agencies are becoming tainted by commercialization as universities and individuals try to cash in on the process. The drying up of government funding of research intensifies the impact of industrial sources of funding focused on short-term (1–2 year) applied research. The question remains: Should taxpayers be paying to allow a few people to profit from an idea developed at public expense, even if the product is of general benefit to society?

Ecological Activists

Although the initial worries about the rampant spread of genetically modified organisms from laboratories were unrealized, some ecologically concerned scientists and citizens feel that the potential for disturbing the balance of natural systems is actually greater today than in the 1970s. During the moratorium period of 1975–1977, a series of experiments were carried out to evaluate the risk of exchanging engineered genetic material between organisms of different types using "crippled" strains of host bacteria. The restrictions on the types of genetic experiments that could be performed were gradually relaxed. Small-scale controlled releases of genetically modified bacteria to colonize host plants and pests were carried out to estimate the risk of perturbing the local ecosystem. Modified plants containing biopesticide or other pest-resistance genes and herbicide-resistance were also tested in small field plots under a variety of conditions. Although such experiments provided little evidence that genetic escape occurred or that pests developed resistance, many people felt that more extensive testing would be required to evaluate risk properly. The early tests were often carried out under duress, after litigation by groups such as the Foundation on Economic Trends. Detailed study of the risk issues was not possible under those conditions. Small test plot sites for modified plants now number in the hundreds, but sabotage of experiments and destruction of valuable research stocks by activists dissuade companies from doing larger, more informative testing. There remain nagging questions about the effects of large-scale exposure of the organisms to the environment—millions of acres of monoculture in different environments with different species of plants and pests in contact with the crop. There are similar worries about the cultivation of fish and other aquatic or marine organisms made transgenic for growth hormones or other traits to enhance growth and how escapees might outcompete the wild populations.

The quietly growing consolidation of agriculture biotechnology under the control of major chemical companies has been noted with concern by Steve A. Edwards, Ph.D. (*Genetic Engineering News* 17[19]: 1, 1997) who observed the recurring vertical business integration of chemicals, seeds, genetic engineering, and pharmaceuticals. Twenty percent of the stock of the last major seed company (Pioneer Hi-Bred) has recently been acquired by Dupont. Examples of large chemical companies and their whole- or partly owned subsidiaries include Monsanto, Calgene, Agracetus, Ecogen, DeKalb Genetics, Holden's Foundation Seeds, Corn States International, Asgrow Agronomics, Monsoy (Brazil); Novartis (Horthrup King, S&G Seeds, Ciba Seeds); Rhone-Poulenc, and the Dow Elanco cooperative venture between Dow Chemical Company and the pharmaceutical company Eli Lilly (Mycogen, DNA Plant Technology, Empresa La Moderna [Mexico]). Besides ensuring a market for their herbicide chemicals with resistant plants, the acquired seed companies provide local outlets for products to generally conservative farmers and access to the seed company's germplasm of multiple crop varieties developed through traditional genetic means.

Widespread use of recombinant microorganisms in bioremediation in which the modified bacteria or fungi are directly released into the environment is severely hampered by the lack of knowledge about microbial ecology. Cleaning up oil spills or contaminated soil and water or biomining metal ores and oil and natural gas recovery are socially and economically valuable uses of genetic engineering. Unfortunately, they are also encumbered by large risk factors because of the extensive exposure of the organisms to the environment.

Politicians

The intrusion of genetic engineering into the political scene has intensified after a hiatus lasting from the middle 1970s. The upswing of political interest has been fueled by vocal public demands for safeguards against decisions being made based on genetic information that to many people felt like invasion of privacy and, like gender and race, was beyond their control. Concern about personal privacy and genetic information has resulted in both federal and state legislation in the 1990s designed to restrict access to and use of genetic information in employment and insurance decisions. At the same time, the use of genetic information and the establishment of databases for forensic purposes have expanded. The issue of database access will have to be addressed as more and more genes are discovered and cryptic DNA markers lose their anonymity. Guidelines, the rudiments of which are addressed in recently proposed "Genetic Confidentiality" legislation (H.R. 306, 341, and 1815 and S. 89 and 422; 105th Congress), will be required for the appropriate use of genetic information and proper communication of test results, consequences, and healthcare choices to clients.

The issue of mammalian somatic cell nuclear transfer (reproductive cloning) has touched even deeper chords of philosophical and religious faith because it addresses the deeply felt if only poorly understood notion of an individual and the relationship of genetics to that view. Likewise, for embryonic stem cells, ethical issues surrounding destruction of embryos touch strongly held views on abortion. Although it is too early to tell, the flurry of proposed broad legislation will likely give way in both areas to carefully circumscribed restrictions on certain kinds of human cell manipulations while allowing other mutually agreed upon medically beneficial work to proceed.

In the future, another round of legislative activity can be anticipated when gene therapy finally comes to fruition as a safe and effective means of treating certain disease states. A series of issues including consent for the procedure, protection against discrimination, and germ line genetic modification will need to be considered.

References

Bullard, Linda. 1987. "Killing Us Softly: Toward a Feminist Analysis of Genetic Engineering." *Made to Order.* New York: Pergamon Press (from Columbia World of Quotations, CD-ROM).

Evenson, R. E., V. Santaniello, and D. Zilberman, eds. 2002. *Economic and Social Issues in Agricultural Biotechnology.* New York: CABI Publishing.

Haseltine, W. A. 1997. "Discovering Genes for New Medicines." *Scientific American* (March): 92–97.

Lamb, G. M. 2004. "Beyond the Mirage of Cell Science." *Christian Science Monitor* (September 16): 16.

Nottingham, S. 2002. *Genescapes: The Ecology of Genetic Engineering.* New York: Zed Books/St. Martin's Press.

Nussbaum, M., and Sustein, C., eds. 1998. *Clones and Clones: Facts and Fantasies about Human Cloning.* New York: W. W. Norton.

Rifkin, J. 1998. *The Biotech Century: Harnessing the Gene and Remaking the World.* New York: Jeremy P. Tarch/Putnam.

Sahai, S. 1997. "Developing Countries Must Balance the Ethics of Biotechnology against the Ethics of Poverty." *Genetic Engineering News* 17(10): 4.

Swanson, T., ed. 2002. *Biotechnology, Agriculture, and the Developing World: The Distributional Implications of Technological Change.* Northampton, MA: Edward Elgar.

Wilmut, I., A. F. Schieke, J. McWhir, A. J. Kind, and K. H. S. Campbell. 1997. "Viable Offspring Derived from Fetal and Adult Mammalian Cells." *Nature* 385: 810–813.

3

Worldwide Perspective

International Impact of Genetic Engineering

Although the United States and other developed nations took the lead in creating genetic engineering technology and then pursuing its applications, in the 21st century many nations have launched their own versions of the industry model. It surprises many Americans that the major international concern is not gene therapy, reproductive cloning, or stem cells, but food and the environment. The greatest controversy stems from the expanding role that multinational agribusiness is playing in agriculture and food production and how the spread of genetic engineering technology is increasing the corporate grip on the global food supply. This is particularly contentious when the technology is exported to nations with a milder regulatory climate or lax enforcement of environmental restrictions. A number of Asian nations outside of Japan fall into this category. The economic potential for industry in these nations, many of which provide significant governmental support, overrides public concern about safety or environmental damage for those in power.

The antipathy toward genetic engineering as an exotic technology is strongest in Europe. For better or for worse, this has led to a measured response to the introduction of the new technologies and concern about their impact, erring on the side of regulatory control. There is a movement to induce countries without strict controls to be responsible and enact minimal standards to avoid blatantly unsafe practices. Releases into the environment do not respect national boundaries. The high-yield agriculture that feeds the world's population rests on dangerously few cultivars of a small number of crops. Disruption of this system would have serious repercussions for all nations.

Status of Genetic Engineering around the World

Because of the technological complexity of genetic engineering, there is a tendency to pay attention only to its impact on those nations with the scientific and technological base to support their own programs in that area. The economic interdependence of almost all of the world's political subdivisions, however, links them in a web of supply and demand for raw materials, goods, and services. Where a given country's economic base fits within this structure depends on its ability to contribute to both sides. This, in turn, depends on the natural resources of that country, including the usual land, forests, water, minerals, and finally the labor pool that extracts those resources and converts them to a useable form. Based on the relative proportion of their economy devoted to the production of goods and services compared with that required to feed the population, nations are classified as "developed," "developing," and "underdeveloped" or "emerging." The impact of technological change can be quite different depending on that nation's economic stage of development. Admittedly Western in nature, with the collapse of the socialist approach with the former Soviet Union, this paradigm currently dominates the world economic structure.

Considering the limited investment of the developing and underdeveloped nations in the educational and technological resources required for the development of substantial biotechnology programs of their own, it is feared that these countries may be destined to be net consumers of genetic engineering and biotechnology, at least in the near term. A restricted industrial base combined with a historical role in supplying raw or minimally processed materials to the more developed nations further constricts their options.

Although some people feel that these restrictions are artificially, and therefore unfairly, imposed by the developed nation's "club," it is also true that nations such as Japan with limited natural resources in the form of raw materials have shifted their economies to manufacturing and services by making the required investments in education and technology. The situation is different for each nation, and the same options are not available to all because of the natural situation or historical developments from past commitments. In addition, the conflict is often drawn between rich and poor nations, which adds an ethical dimension to the otherwise sterile consideration of economic and technical questions.

A number of international organizations already concerned with problems of economic development have begun considering the effects of biotechnology on the relationships between the developed and other nations. These include the World Health Organization; the International Council of Scientific Unions; the United Nations (UN) Educational, Scientific, and Cultural Organization; the Food and Agriculture Organization; the United Nations Industrial Development Organization; and the United Nations Development Program. They have the difficult job of attempting to reconcile the divergent interests of competing nations to reduce the economic dislocation and social disruption fueled by international trading forces (Swanson 2002). With the proper set of guidelines adhered to by all parties and negotiations in good faith, it should be possible to minimize further dependency of the underdeveloped and developing nations on the developed nations while ensuring supplies of raw materials and reasonable markets for finished products for the manufacturing economies. Whether this will come about remains to be seen. Neither the form these guidelines should take nor the forum for producing and enforcing any such agreements has been established. The following sections detail the issues at the heart of the discussion and the present framework for dealing with the problems.

Impact on Developing Countries

The Green Revolution sparked by traditional plant breeding and application of high-production principles and methods provides lessons for what might happen with genetic engineering applied to biotechnology. In many places during the 1960s, veritable miracles were achieved. The yields of cereal crops in parts of the third world (today's underdeveloped nations) doubled or tripled between 1964 and 1988 with the introduction of the improved plant varieties and cultivation methods. In the Indian Punjab, wheat and rice harvests were improved as were sorghum, finger millet, and cassava in rain forest areas. By contrast, in areas relatively untouched by the Green Revolution such as Africa, poverty increased. In developing and particularly in underdeveloped countries, a growing majority of the world's poor depends primarily on wage employment, not on income from crops raised on farmland. The Gene Revolution of biotechnology, agribusinesses are quick to point out, similar to the Green Revolution, has the potential to relieve the plight of poor people by providing greater and more stable employment, better nutrition, and increased small farm income (Thomson 2002). At the same time, these technologies stand to further harm the poor by displacing labor or otherwise improving the competitiveness of low-labor large farms to the exclusion of small farms and their workers. A similar scenario exists in developed countries such as the United States where the family farm is in danger of extinction in some areas. The displaced workers in the United States, however, have considerably better economic fallback options than those in the third world.

One of the criticisms of the use of present biotechnology applications in developing or underdeveloped nations is they are directed solely at developed nation marketing demands. These include obtaining staple crops with improved health benefits, by, for instance, lowering the erucic acid content or, as in the Philippines, improving yellow maize varieties to feed chickens. Unfortunately, in the Philippine case, the amount of white maize, a staple food of the country's poor, decreased over that same time. The great majority of the world's undernourished requires more calories; only a minority requires more protein. Shifting land from staple food to cattle raising or cattle feed production will reduce the calories available. For both technical reasons and because the genetic engineering capacity is concentrated in the northern, developed countries with different interests, biotechnological or even traditional plant-breeding improvements have been slow in coming to tropical staple plant crops such as cereals. Propagation of valuable plants though tissue culture of plant clones, bypassing genetic engineering per se, has been widely applied instead to coconut, palm, potatoes, coffee, and tea. This is the type of biotechnology research most undertaken in developing countries.

A concern of developing and underdeveloped nations is that the biotechnology products developed from genetic engineering represent yet another form of economic enslavement to the developed world. The business plan is doing little to rebut this conception. With the current generation of genetically engineered products, the benefits of the technology remain with the producer. The pesticide- or herbicide-resistant plants or F1 hybrid seeds have to be purchased fresh each year from the supplier in the developed country. Second- and third-generation genetic technologies are being implemented to prevent farmers from saving seed from a harvest to plant the next season circumventing a 50-year tradition in Europe of Plant Breeders Rights to re-plant seed they have grown. Eighty percent of farmers in developing countries plant retained seed. Resource-poor farmers practice 60% of global agriculture and produce 10% to 15% of the food (Swanson 2002). Other technologies under development will require application of special agents to a crop to take advantage of the bioengineered enhancements. A growing merger trend and chemical companies buying up agricultural biotechnology and seed companies are disturbing to many because it gives these companies, which are frequently multinational, the ability to guarantee a market for their own herbicides and other products.

Major political and economic philosophies divide national interests, which also take on ethical overtones. The Green Revolution was researched and developed to a large extent by the public sector in government facilities and through government-sponsored research grants to universities and foundations, consistent with the expanded role of government in society at that time (Evenson, Santaniello, and Zilberman 2002). The present political environment, particularly in the United States where much genetic engineering research is being done, favors the private sector in which funding for research and development is provided by private investors. The legislative environment in the United States at the same time supports the issuance of patents protecting property rights to organisms and processes central to genetic engineering and biotechnology. The combination has resulted in the escalation of private, for-profit competitive industry concerned with economic survival at the expense of the public sector, which, in theory, can take a longer view of issues. A number of factors converge to produce a monopolistic situation with all of the key supplies in the rich world. Only a few of the underdeveloped and developing countries (Brazil, China, India, Mexico, Pakistan, Cuba, and some others) possess the resources, scale, capacity, or stable political environment to mount a significant public-sector biotechnology effort on their own.

Another concern of some developing nations as well as environmentalists worldwide is that genetically engineered organisms—in most cases, plants—will be released on a large scale into the environment without sufficient relevant testing to ensure that they will not spread to devastate local ecosystems. The anxiety is based on the same questions about safety that have been discussed in chapter 2. There is a temptation for companies to market or to test crops initially in countries without stringent regulations to speed up product development or to avoid expensive trials. Environmentalists fear that debt-burdened or corrupt governments may succumb to the lure of promised quick-and-easy profits for their country (or private pockets) to introduce a technology before it is proven safe. In some cases, these are the same governments that allow the use of polluting pesticides and herbicides because they are inexpensive that the rest of the world has banned. These governments could be mortgaging their country's future for today's profits and at the same time promoting ecological disaster.

It is in the best interest of the developing nations to work out some kind of plan to equalize the impact of genetic engineering on their economic dependence on the developing nations. They should do so sooner rather than later. In the developed countries, given sufficient economic incentive, biotechnology can produce substitute products for developing nation exports. This scenario would have catastrophic consequences for the largely undiversified economies of developing nations. Escagenetics (California) is working on a commercial-scale tissue culture method to produce vanilla plantlets bent on capturing the $200 million annual U.S. market presently dominated by Madagascar. Nestle's (the Netherlands) and the Kao Corporation (Japan) are employing genetic engineering to make cocoa butter substitutes. Cacao represents the second most important agricultural commodity in the third world. Africa accounts for nearly 60% of world production of this crop. A German biotech firm is working on a coffee substitute. The third world is responsible for virtually all coffee production. The United States alone imports $10 to $50 billion of coffee yearly from such countries as Columbia, Burundi, Uganda, Rwanda, Ethiopia, and Indonesia.

What leverage do developing countries possess to avoid economic indentureship to rich developed nations? While insisting on the adoption of uniform patent protection for biotechnology products, the developed nations assert a claim to free access to the genetic resources of the plants and animals of the developing and undeveloped nations as a world right. By some estimates, Asia, Africa, and Latin America have provided the genetic base for 95.7% of the world's food crops. In comparison, Europe, North America, and Japan, comprising the developed nations, contributed only 4.3% of the diversity. The developing nations argue that the "wild" germplasm has been modified by centuries of cultivation, breeding, and modification and that it should not be a costless resource to be exploited by companies for products that are subject to patent and trade secret and then sold back at premium prices to the countries providing the original resource. A UN Food and Agriculture Organization proposal to establish world Gene Banking Centers with free access to all countries was rejected by the United States, the United Kingdom, Germany, France, Denmark, Norway, Sweden, Finland, the Netherlands, and New Zealand. The U.S. Gene Bank at Fort Collins, Colorado, although U.S. government property, has refused grants of germplasm to countries with which the United States had political disputes, such as the former Soviet Union, Cuba, Afghanistan, Albania, Iran, Libya, and Nicaragua. The Treaty for the Protection of Biodiversity signed at a UN Conference on Environment and Development held June 1–12, 1992, in Rio de Janeiro by 153 countries was not signed by the United States until 1993 because the treaty did not incorporate intellectual property protections. Instead, it contained language dealing with regulation of biotechnology, equitable sharing of research and development benefits and technology transfer, national registration of biological resources, rights of indigenous peoples to profits from their own plants and knowledge, and prior informed consent to any biotech testing in their country. The World Trade Organization attempted to address the commercial issues with the Trade Related Aspects of Intellectual Property Rights agreement signed in 1995. These commercial and ethical issues remain to be reconciled.

So what help might genetic engineering bring to the poor of underdeveloped countries? Increasing the output of cash crops or reducing imports with disease-resistant and drought-tolerant nitrogen-fixing cotton, for example, without lowering employment would be an advantage. This engineering would need to be applied judiciously to products that are labor-intensive such as certain vegetable oils, but not to tea, coffee, or palm oil where a price glut would be devastating to the economy. Genetic engineering can reduce the need for expensive or imported fertilizers, a concern that was exacerbated by the Green Revolution. Improvement of the shelf life of products by, for example, changing the oil composition in a foodstuff to slow spoilage would help poor countries utilize a higher proportion of their harvest as they are the most likely to have inadequate marketing capacity or infrastructure for processing.

Possible Impact of Genetic Engineering

Developed nations seldom consider publicly the economic impact of widespread use of genetically engineered organisms on nations that would be competitors for raw materials or that would be net importers of the technology. Whether such repercussions on a given economy are severe is not likely to halt the application of genetic engineering, given the strength and momentum of world economic forces (Bernaer 2003). However, the resultant economic and political dislocation can be moderated by appropriately sensitive policy implementation. It would be in the developed nations' best economic interest to avoid creating a dependency situation with frustrated and desperate governments in countries where the major export is replaced or reduced in value and food cannot be raised in sufficient quantities for the people because the arable land is growing export crops. For both technical and economic reasons, the food crops for the mostly tropical underdeveloped or developing nations are not the types of plants being genetically engineered today. Requiring producers to buy high-tech seeds every year from the developed countries who then control the price on the product in their market has been a problem with hybrid strains in the past, and genetically engineered strains, especially the new "terminator" strains that cannot be replanted, will only make it worse. Enlightened self-interest would seem to dictate some thought about balancing the economic consequences against the initial gains. Unfortunately, such action is rarely seen in the real world unless an issue is made of it. The developed nations and multinational agribusinesses again seem inclined to pursue their own interests as evidenced by their strict insistence on protecting the full intellectual property rights operant in Western economic systems while devaluing genetic diversity contributed by the developing and undeveloped nations' plant and animal resources.

Recommendations for Action

An issue often not commented on is the economic impact of the new genetic technologies on the developing and underdeveloped nations of the world. Genetic engineering is a highly technical field requiring a substantial investment in both training and technology to be able to do the research and to meet regulatory requirements. It is no accident that the developed nations, including a few of the developing nations, have cornered the market on medical applications and food production. Developed nations become the sole suppliers of patented engineered seeds and plants, which must be purchased annually with proceeds from the previous year's harvest of products in demand by the developed nations. Consequences of this include effective national economic servitude and decreases in indigenous food crop production to plant enough imported seed for export of products. In response, there are disputes, such as those aired at the 1992 Rio de Janeiro Conference on Biodiversity, over intellectual property rights and the value of the contribution of genetic diversity provided by the undeveloped or underdeveloped nations to the forerunners of the engineered plants. A compromise needs to be reached because the developed nations could, if sufficiently pressed, eventually produce synthetic or engineered replacements for the major exports of the underdeveloped nations, triggering an economic disaster for suppliers.

Biowarfare and Bioterrorism

One internationally relevant area in which there is little information available, largely as a result of its classified nature, is the military use of genetic engineering to provide offensive weapons. Despite a treaty renouncing such research, it is difficult to ascertain exactly what the situation is because all parties deny involvement in offensive weapons work. Concern over Iraq's capacity to manufacture and deploy biological weapons of mass destruction was a primary justification for the United States to invade that country in March 2003.

In the aftermath of the September 11, 2001, attacks on the World Trade Center and Pentagon, the prospect of bioterrorism was given new credence by a series of anthrax-laced letters sent through the postal system. Although the source of the weaponized material remains hazy and those responsible have not been apprehended, the use of such agents as weapons of fear has been established. Biotechnology has been enlisted in counter-bioterrorism programs designed to provide rapid detection and identification of an attack. Other programs are aimed at protecting the water and food supply. Such technology is a double-edged sword because it can be used to engineer terror weapons.

International Mirror of Controversies in the United States

In general, the issues involving genetic engineering that are receiving the most attention are the same in the United States and the rest of the developed world. The difference is in the relative importance placed on the various issues. Genetically modified food heads the list abroad, followed by environmental genetic contamination. Lower down on the scale are issues such as human reproductive cloning, stem cells, and genetic testing, which rank near the top in the United States. This is not to say that there hasn't been debate or movements to regulate the use of genetic engineering technology, but the relative importance to the citizens of different countries is notable. What an American might think is the most crucial point might be of relatively minor importance to a European. This leads to misunderstandings and international angst.

Genetically Modified Foods

There is considerable resistance in the United States to the widespread use of biotechnology in the form of genetically modified food plants and animals generally referred to as GMOs (genetically modified organisms). This is despite the fact that the United States has the largest acreage of GMOs under cultivation in the world. Four countries grew 99% of the developed world's transgenic crops in 2000, with the United States accounting for 68% of that total. Developing nations now account for one-third of worldwide GMO crop acreage. The technology is so widespread that the question in the United States is whether foods derived from GMOs should be specifically labeled, a step that has been vigorously opposed by agribusiness as potentially stigmatizing products for which there is no proof of harm. A compromise has been the establishment of certification for "non-GMO" labeling, akin to the earlier "organic" labeling for products produced without herbicides, pesticides, or synthetic fertilizers.

Particularly in Europe and Japan, GMO-derived products may not be produced or imported. In October 1999, the European Commission's Standing Committee for food established a legal threshold of 1% accidental GMO contamination for rejection. Even transportation in containers that have been used to carry GMO-derived products has been challenged, and testing for the modified genes in products is routine. This antipathy is of concern to U.S. farmers because the policies restrict their potential markets. Even if a farmer plants only traditional cultivars, contamination from a neighbor's field or an unwashed storage bin or railcar somewhere along the way to market could make their crop worthless.

International Regulation of Genetic Engineering

Regulatory approval for products derived from genetic engineering depends greatly on the country and on the product being considered. Biopharmaceuticals have long been regulated, and the advent of materials produced through genetic engineering required relatively minor changes in the testing and approval process for most pharmaceutical products. Organizations dealing with GMOs internationally include the National Institute for Biological Standards and Control (United Kingdom), UN Conference on Environment and Development (multinational), Biotechnology Quality Work Party (European Union), Center for Biologic Evaluation and Research (United States), Advisory Committee on Novel Foods and Processes (United Kingdom), and the International Conference on Harmonization (multinational).

Agribiotechnology has had a considerably less positive experience. Part of the reason for this is that modifications of current GMOs benefit the producers (herbicide resistance, plant viral resistance, insect resistance). Consumers feel that they and the environment are put at possible risk with no benefit to them. The introduction of several nutritionally enhanced crops such as vitamin A–enriched rice or low *trans*-fatty acid canola oil has had little impact on this opinion. The public is much quicker to appreciate the production of scarce medicines that are not available from other sources.

References

Bernaer, T. 2003. *Genes, Trade, and Regulation: The Seeds of Conflict in Food Biotechnology.* Princeton, NJ: Princeton University Press.

Evenson, R. E., V. Santaniello, and D. Zilberman, eds. 2002. *Economic and Social Issues in Agricultural Biotechnology.* New York: CABI.

Swanson, T., ed. 2002. *Biotechnology, Agriculture, and the Developing World: The Distributional Implications of Technological Change.* Northampton, MA: Edward Elgar.

Thomson, J. A. 2002. "The Potential of Plant Biotechnology for Developing Countries." In *Biotechnology and Safety Assessment,* edited by J. A. Thomas and R. L. Fuchs. San Diego, CA: Academic Press.

4

Chronology of Genetic Engineering

Although the high-technology aspects of genetic engineering have come to public attention mainly since the recombinant DNA revolution in the 1970s, similar activities have actually taken place over much of the thousands of years of human history. For much of that period, people were searching for the order in nature to explain how phenomena related to one another. Only in the latter part of the 19th and 20th centuries did knowledge of genetic processes progress to the cellular and molecular levels to the extent that scientists now feel that they understand much of the detail. Using this insight, they have begun to manipulate the systems in a variety of ways. The following chronology highlights key developments that led to today's understanding of how genetics works and events that were important in society's reaction to science's new capabilities.

6000?
B.C. Alcoholic beverages, cheese, and other products are produced by action of naturally occurring yeast present in foodstuffs by the Sumerians and Babylonians in Mesopotamia. Around 4000 B.C., the Egyptians discover how to make leavened bread using yeast. Additional fermentation processes are in use in China, including wine making in the Shang Dynasty around 1300 B.C.

1790 The United States passes the first patent law to provide commercial protection for inventors.

1804   After much discussion and experimentation, chemists conclude that substances consist of identical invisible particles called *molecules* that are themselves formed from a smaller number of more elementary particles that John Dalton called atoms. An early theory of the atomic nature of matter had been stated by the Greek philosopher Democritus (460?–352? B.C.).

1810   The chemical equation for alcoholic fermentation $[C_6H_{12}O_6 = 2\ CO_2 + 2\ CH_3CH_2OH]$ (sugar converted to alcohol and carbon dioxide) is deduced by the French chemist Joseph Louis Gay-Lussac, explaining a process that had been in use for millennia.

1836   The German physiologist Theodor Schwann attributes processes of putrefaction and fermentation to microorganisms. In the 1860s, the French chemist and microbiologist Louis Pasteur (1822–1895) will provide the experimental evidence for this. Such explanations do not conform to popular sentiment.

1852   Sparrows are imported from Germany into the United States to deal with caterpillars ravaging food crops. This example of the spread of a foreign species in a new environment is thought by some to show what could happen with a new genetically engineered plant or animal carrying foreign genes that give it an advantage.

1862   The Organic Act establishes the U.S. Department of Agriculture, removing it from the Patent Office and directing the department to collect and distribute seeds and plants to farmers.

1865   In a presentation before the Natural Science Society in Brunn, Austria, an Augustinian monk from Brno, Gregory Mendel (1822–1884), proposes that invisible units called *factors* are passed from one generation to the next and that they account for transmission of observable traits. Ignored in the wake of the furor over Charles Darwin's evolutionary theory introduced in 1859, Mendel's publication is independently rediscovered by the botanists Hugo DeVries (France), Erich Von Tschermak-

Seysenegg (Austria), and Carl Correns (Germany) in 1900.

Louis Pasteur develops the germ theory of disease from his work on the silkworm disease plaguing the French silk industry. The English surgeon Joseph Lister begins using disinfectants such as carbolic acid (phenol) during surgical operation in the handling of patients and in the care of wounds, dramatically reducing patient deaths from infection.

1869   The Swiss biologist Johann Friedrich Miescher isolates DNA from white blood cells in the pus obtained from discarded bandages. Because he is busy investigating the nature of the many chemicals found in cells, he fails to make the connection between DNA and Mendel's hereditary "factors."

1877   Louis Pasteur notes that some bacteria die when cultured with certain other bacteria. He suggests that some bacteria produce substances to kill other bacteria but goes no further with the concept.

1882   The German bacteriologist Robert Koch first determines the cause of a human microbial disease with the identification of the tuberculosis organism. In the process, he establishes that specific diseases are caused by specific organisms by isolating pure cultures of bacteria from infected individuals. Similar observations and culture techniques are used by Pasteur to produce vaccines against anthrax and rabies.

1885   Although the Chinese have practiced a desensitization treatment for centuries and the English physician Edward Jenner developed a smallpox vaccine in 1796, the presence of disease-causing microorganisms had not been determined. Vaccination becomes an acknowledged weapon in the medical arsenal when nine-year-old Alsatian Joseph Meister is successfully treated for rabies by Louis Pasteur.

1888   Discovered in 1879 by the German biologist Walter Flemming, another German biologist, Heinrich Wilhelm

1888 Gottfried Waldeyer, uses the word *chromosome* to de-
(*cont.*) scribe the staining of structures in the cell nucleus by certain colored dyes. It would be 14 years before the connection is made to the chromosomal role as a repository
for genetic information.

1889 The vedelia beetle, commonly known as the ladybug, is
imported from Australia into California to control the
cotton scale insect that is devastating the state's citrus or-
chards.

1890 The discovery of antibodies by the German bacteriologist Emil Adolph von Behring provides part of the explanation of the protective factors that vaccines elicit.
The special cells that carry out the rest of the functions
necessary for immune system surveillance of foreign invaders are described by the Russian zoologist and bacteriologist Elie Metchnikoff around 1891.

1892 Viruses are described by the Russian bacteriologist
Dmitri Iosifovich Ivanovski as disease-causing agents
smaller than bacteria.

1893 The German Nobel Prize–winning chemist Wilhelm Ost-
wald proves the catalytic nature of protein components
of cells called enzymes that are responsible for cellular
chemical reactions such as the fermentation of glucose.
One enzyme molecule causes the conversion of many
substrate molecules into product. These biological catalysts greatly speed up the rate of cellular reactions that
occur much more slowly without them.

1895 A German company, Hochst am Main, sells commercially
cultured *Rhizobium* isolated from roots nodules to enhance
soil nitrogen fixation in Europe. These soil bacteria are introduced into the United States in the following year.

1899 Chemical modification of the salicylic acid extracted
from willow bark and other plant extracts used to combat fever yields acetylsalicylic acid, which causes less
stomach irritation. This first "wonder drug" is marketed
as Aspirin® by Bayer, a German chemical company.

1901 The chemist Franz Hofmeister advances the theory that the vital reactions of life are performed by enzymes, those catalytic molecules that make chemical reactions efficient enough to power life processes. Direct proof for this comes when mutations (genetic changes) crippling enzyme activities are shown to be injurious or lethal for bacteria, fruit flies, and people.

1902 The German chemists Emil Fischer and Franz Hofmeister show that proteins are strings of individual amino acid units chemically attached end to end through peptide bonds to give polypeptides. This theme of using the same chemical reaction to link varieties of a common building block into long chains or polymers to obtain desirable properties is repeated in biology with DNA, RNA, and polysaccharides.

The U.S. cytologist and, later, physician Walter Stanborough Sutton suggests that Mendel's "factors" reside on chromosomes and are segregated as the chromosome pairs separate during the formation of gametes through meiosis. He names these factors *genes*, the term we use today.

1903 The first use of the term *biochemistry* to describe the chemistry of life is attributed to chemist Carl Neuberg.

1905 A system for analyzing the occurrence of observable traits (phenotypes) describing the genetic principles of linkage and gene interaction is used by William Bateson and Reginald Crundall Punnett. These pioneering English geneticists coin such genetic terms as *F1 and F2 generations, allelomorphism, homozygote,* and *heterozygote* that are used to describe genetic populations of individuals. The molecular events underlying the observations of the inheritance of traits are eventually worked out over a half century later.

1907 U.S. biologist Ross Granville Harrison establishes the culture of isolated cells or tissue separate from the intact organism (in vitro). Refinements in these techniques have allowed in vitro culture of many types of cells, a

1907   boon to the study of cellular biochemistry. The tech-
(*cont.*)   niques are eventually extended to tissue and cellular
transplants, monoclonal antibody production, plant
propagation, and genetic engineering modification of intact organisms from the 1970s onward.

1908   The English physician Sir Archibald Edward Garrod recognizes that the final product of a gene is a protein and
pioneers studies of genetic diseases in humans with his
work on alkaptonuria. His postulate that diseases can be
caused by mutant genes is unnoticed until it is rediscovered in 1940.

1909   Wilhelm Johannsen, a Danish botanist, first uses *phenotype* and *genotype* to describe an observable trait and the
genetic factors (genes) responsible for the trait, respectively. Johannsen also describes the process of "selection" for a genetic trait by which an advantage conferred
by a gene leads to a higher occurrence of that gene in the
next generation because it enhances survival.

1910   American Thomas Hunt Morgan, a Nobel Prize–winning
geneticist, establishes the basis of modern genetics by
proving that genes reside on chromosomes, pinpointing
genes to particular chromosomes of the fruit fly,
Drosophila melanogaster. Over the next 10 years, Morgan's
group describes gender-linked genes (white eye color in
Drosophila) and other trait linkages. Morgan and col-
leagues publish the classic *The Mechanism of Mendelian
Heredity* in 1915, formally setting forth the principles of
gene theory and linkage.

1912   Butanol and acetone are produced industrially via microbial action using a process developed by the American
Chaim Weizman, who later becomes president of Israel.
This is the first large-scale use of microbial processes for
products other than food.

1917   A virus that infects and destroys bacteria, a *bacteriophage*,
is independently discovered by the English bacteriologist Frederick Twort and French bacteriologist Félix Hu-
bert d'Herelle. Bacteriophages, such as the T4 and the

lambda phage, are important in early studies of gene structure by the U.S. physicist-turned-geneticist Seymour Benzer and others in the 1960s. The lambda phage has been used as a gene-cloning vehicle since the 1970s.

1926   The first crystallization of an enzyme, urease, by James Batcheller Sumner proves it to be a protein. He is awarded the Nobel Prize in Chemistry in 1946 for this achievement.

Henry Agard Wallace founds the Hi-Bred Company, a hybrid corn seed producer, known today as Pioneer Hi-Bred International, Inc., a company involved in genetically engineering agriculturally useful plants.

1927   The first localization of a specific gene to a particular chromosome in mammals is accomplished by the geneticist Theophilius Shickel Painter, who correlates a visible chromosomal deficiency (partial deletion) in a mouse with a genetic analysis.

1928   Frederick Griffith demonstrates that a transforming principle (shown to be DNA in 1944) in *Pneumococci* converts nonvirulent to virulent strains—a "natural" (unassisted by humans) form of gene transfer.

English bacteriologist and Nobel Prize winner Sir Alexander Fleming discovers the first clinically useful antibiotic, penicillin, which is produced by a mold growing on a contaminated bacterial culture plate. Other antibiotics are known at this time but are toxic to mammalian hosts. Penicillin is isolated in 1938 by chemists Howard Florey and Ernst Chain of Oxford University in England. Although industrial production is initially delayed by lack of patent protection, the antibiotic is eventually made available for clinical use some 15 years later.

1932   A primitive microscope using electrons for sample illumination is built by M. Kroll and Ernst August Friedrich Ruska. High energy, and thus short wavelength, electrons allow visualization of DNA and single protein molecules, objects some 1,000 times smaller than can be seen with a regular light microscope. Being able to see what

1932
(*cont.*)
previously could only be imagined allows scientists to work out how molecules are assembled to form the parts of a cell.

1935
A. G. Tansley introduces the concept of an ecosystem as a balanced interdependent network of relationships linking organisms and their environment. The minimizing of competition by filling compatible "niches" is believed to be critical for the balance and functioning of ecosystems. Changes such as the introduction of a new species or an old species that develops a novel competitive advantage can upset the balance with potential catastrophic consequences. Today, opponents of the release of genetically modified organisms worry that precisely just such a scenario will develop.

1938
X-ray scattering is first used to study the folded structure of DNA by physicists William Thomas Astbury and F. O. Bell. This technique, pioneered by English physicist Sir William Lawrence Bragg in 1912 for small molecules, also proves suitable to determine the structure on the scale of the atom of even large DNA and protein molecules.

Bacillus papilliae, a bacterium, becomes the first microbial product registered by the U.S. government as a control for Japanese beetles. Large-scale spraying over several states on the East Coast is the first substantial release of such microbes.

1939
French bacteriologist René Jules Dubos isolates the antibiotic gramicidin from a common soil bacterium. Dubos later becomes a noted environmentalist.

1940
U.S. geneticist George Wells Beadle and U.S. biochemist Edward Lawrie Tatum, cowinners of the Nobel Prize in 1958, propose that one gene specifies one enzyme from studying the inheritance of traits in the common bread mold, *Neurospora crassa.*

1941
The term *antibiotic* is coined by Ukrainian-born U.S. microbiologist and Nobel Prize winner Selman Abraham Waksman to describe compounds produced by microor-

ganisms that kill bacteria to gain an advantage in the competition for nutrients and space. These toxins are designed to be very specific against the bacteria because the microorganisms producing them have to live in the same environment. Pasteur had made the original observations of "antibiosis" in 1877.

Although publicly denouncing germ warfare, President Franklin Delano Roosevelt approves a secret plan to develop a U.S. biological warfare capability. By 1942, the United States has a 4-pound anthrax bomb.

1944   DNA is recognized as hereditary material when microbiologists Oswald Avery, Colin MacLeod, and Maclyn McCarty transmit the virulence of a strain of *Pneumococcus* to a nonvirulent strain with isolated DNA alone.

1946   C. Auerbach and J. M. Robson observe gene mutation by chemicals with subsequent alteration of traits. These observations provide the rationale for explaining genetic diseases.

Transfer of DNA and genetic traits between *Escherichia coli* strains by conjugation (bacterial mating) is demonstrated by Nobel Prize winners U.S. geneticist Joshua Lederberg and biochemist Edward Lawrie Tatum.

1950   The pairing of purines and pyrimidines across the strands of the DNA helix—adenine (A) with thymine (T) [A=T] and cytosine (C) with guanosine (G) [C=G]—is demonstrated by Erwin Chargaff and coworkers, explaining the constant ratios of these nucleic acids in DNA. This DNA sequence specificity removes one of the last objections to nucleic acid encoding of hereditary information. These findings were instrumental in determining the physical structure of DNA 3 years later.

1951   American geneticist Joshua Lederberg demonstrates that bacteria can exchange part of their genetic material, which he calls a *plasmid*. He also finds that viruses that attack bacteria can act as an intermediate to transfer genetic matter between bacteria. Lederberg is later involved in

1951   the recombinant DNA and biotechnology policy and de-
(*cont.*) bate in the 1970s.

1952 X-ray diffraction patterns of the B form of DNA are described by English physicist Rosalind Franklin.

1953 Using American biochemist Edwin Chargaff's findings
on purine:pyrimidine pairs, Rosalind Franklin's X-ray
scattering data, and hand-built models, the double alpha
helical structure of DNA is described by English biochemist James Dewey Watson, physicist Francis Harry
Compton Crick, and physicist Maurice Wilkins. Rosalind
Franklin dies from cancer shortly after the historical article describing this finding is published.

1955 Using a technique called fine structure mapping, Seymour Benzer, a U.S. physicist turned biologist, shows
that there are many sites within a single gene that are
susceptible to mutation. This meant that there are many
ways a single gene could be altered and that the observed effect might not be the same for all mutations. It
explains why genetic alleles exist and that they are the
same gene, although each maps slightly differently and
sometimes shows a slightly altered trait.

1956 U.S. Nobel laureate biochemist Arthur Kornberg's discovery of DNA polymerase, the enzyme responsible for
the replication of DNA, answers part of the question of
how genetic information is faithfully copied for transmission to the next generation.

The Nobel Prize–winning work of American biochemist
Christian Boehmer Anfinsen reveals that the three-dimensional structure of proteins is determined by the
order of the amino acids in the protein sequence. He
shows that a string of chemically synthesized amino
acids with the same sequence as the biological enzyme
RNase A fold properly and express the catalytic activity
of the native enzyme. Twenty years later, molecular biologists will take advantage of the natural folding of
polypeptide chains when they begin to express large

quantities of nonbacterial proteins in bacteria and then isolate them to study their function.

1957   The remarkable fidelity of DNA replication becomes evident when biochemists Matthew Meselson and Franklin Stahl show that replication proceeds by copying one of the original strands, using the A=T and C=G rules surmised by Erwin Chargaff. The replication proceeds in one direction only from one end (the 5' end) of the DNA, and a "proofreading" activity of the DNA polymerase enzyme checks for mistakes.

1958   Francis Harry Compton Crick and Russian-born U.S. physicist George Gamow propose the "central dogma" of information flow in molecular genetics: DNA codes for RNA, which codes for protein.

Microbiologists Samuel Bernard Weiss, J. Hurwitz, and others report the discovery of DNA-directed RNA polymerase that provides essential support for the hypothesis. Reverse transcriptase, an RNA-dependent DNA polymerase, found in a family of retroviruses that includes some cancer viruses and the HIV virus linked to AIDS, proves an exception to this rule. The independent discovery in 1970 of this enzyme by cell biologists Howard Temin and David Baltimore will prove of immense value to molecular biologists because it allows the cloning of proteins from their expressed mRNA.

1959   An extra chromosome is discovered in the nuclei of cells from Down syndrome children by the French scientists Jerome Jean Louis Marie LeJeune, M. Gautier, and Raymond Alexandre Turpin. This unbalanced extra dose of genes results in mental retardation and a host of other malformations in this genetic disease. This relatively common birth defect occurs spontaneously with increasing frequency when maternal age is greater than 35 years.

French Nobel Prize–winning biochemists François Jacob and Jacques Lucien Monod establish that the internal controls for gene regulation reside in the DNA sequence

1959   on the chromosome as mappable features distinct from
(*cont.*) the portion of genes coding for proteins. The steps in
          protein biosynthesis are also worked out at this time.

1961   The genetic code words for amino acids are identified
          over a period of 5 years in the labs of U.S. Nobel
          Prize–winning biochemists Robert William Holley, Mar-
          shall Warren Nirenberg, Har Gobind Khorana, and
          Severo Ochoa. As stated by the central dogma of molec-
          ular biology, the code is transcribed from DNA into an
          intermediate messenger RNA (mRNA), the function of
          which was earlier defined by Francois Jacob and Jacques
          Lucien Monod. mRNA is subsequently translated using
          the genetic code words for the amino acids into protein.

1963   The first textbook based on the principles of modern
          ecology is published by biologist E. P. Odin.

1964   The Green Revolution begins with the development of
          new strains of rice at the International Rice Institute in
          the Philippines. With sufficient fertilizer, production
          yields of previous strains are doubled. Although hailed
          as providing food for the hungry, issues such as increas-
          ing the dependence of the third world on the developed
          industrial nations for fertilizers and speeding nutrient
          depletion of soils are downplayed. Acknowledgment of
          the impact of these issues will come later with experience
          and will engender significant debate over the introduc-
          tion of genetically engineered crops in the 1990s.

1965   Genes coding for bacterial antibiotic resistance are found
          on *plasmids,* small circular pieces of DNA that remain
          separate from the main DNA of the bacterium. The most
          common antibiotic resistance genes code for enzymes
          that destroy the drug molecules before they can harm the
          bacteria. β-lactamase hydrolyzes β-lactam antibiotics
          such as penicillin.

1967   Polynucleotide ligase from *E. coli* is isolated and studied
          by U.S. biochemists Samuel Bernard Weiss and Charles
          Clifton Richardson. This enzyme is important in DNA

damage repair, closing breaks in DNA by forming a new bond between pieces of the same strand of DNA. A bacteriophage T4 variant of the ligase is a routine tool in engineering recombinant DNA to splice together pieces of genes.

1970   The first restriction endonuclease enzyme, *E. coli* restriction endonuclease I (Eco RI), is isolated. These enzymes, discovered in 1968, protect bacteria in their natural environment from foreign DNA. By cutting at specific nucleotide sequences they make manipulation of DNA predictable and much easier.

1971   General Electric and Indian-born U.S. microbiologist A. M. Chakrabarty apply, initially without success, to the U.S. Patent Office for a patent on oil-eating *Pseudomonas* bacteria, created to deal with ocean oil spills. The patent is eventually issued after review by the Court of Customs and Patent Appeals. This landmark decision removes the distinction between patenting animate and inanimate inventions and opens the infant biotechnology industry to development. In 1980 the U.S. Supreme Court upholds the patent.

1972   The first recombinant DNA molecules constructed using restriction enzymes and DNA ligase are fabricated at Stanford University. In 1973, restriction enzymes are used for cloning toad DNA into bacteria by the laboratories of U.S. biochemists Stanley Norman Cohen at Stanford and Herbert Wayne Boyer at the University of California, San Francisco. A patent issued to Stanford University and the University of California on this process collects royalties today for every cloning experiment performed with these enzymes.

1973   Public concern is expressed over the possible production of dangerous hybrid organisms by the new recombinant DNA technology. Scientists draft a public letter arising from discussions at the Gordon Conference on Nucleic Acids published in *Science* on the possible biohazards of DNA splicing.

1973   Congress creates the Environmental Protection Agency
(*cont.*) (EPA) to act as the national watchdog for the environment.

1974 The National Institutes of Health (NIH) forms the Recombinant DNA Molecule Program Advisory Commit-
tee (RAC) on October 7, which is charged with framing
guidelines for recombinant DNA research and reviewing
gene therapy protocols. Its first meeting is in February
1975. Critics of recombinant DNA technology call for a
worldwide moratorium, which is respected on certain
kinds of experiments while the potential dangers are
studied further.

1975 The first mammalian gene (rabbit globin, the protein part
of the oxygen carrier hemoglobin) is cloned in bacteria
by molecular biologists A. Efstratiadis, F. Kafatos, A.
Maxam, and Tom Maniatis.

The Asilomar Conference on Recombinant DNA Molecule Research is held to discuss concerns and progress in
providing biological containment of experimental organisms and in assessing the safety of recombinant DNA research. Tensions run high because scientists feel that they
are being railroaded by activists who don't understand
or care about scientific issues or the freedom of inquiry.
Nonscientists, on the other hand, feel that the issues at
stake, such as the safety of the environment, are too important to be left for scientists to decide because they
have their own agendas and their livelihoods at stake.
The Senate Subcommittee on Health, Committee on
Labor and Public Welfare, begins the first public debate
on recombinant DNA with a series of hearings on genetic
engineering beginning on April 22 and chaired by Senator Edward Kennedy. Other nations seem to be following
the lead of the United States in the controversy at this
time. The Report of the Working Party on the Experimental Manipulation of the Genetic Composition of Microorganisms in the United Kingdom calls for special
laboratory precautions for recombinant DNA research.

1976 On June 23, the long-awaited publication of the NIH's
first Federal Safety Guidelines on Recombinant DNA Re-

search establishes a voluntary system of regulation for institutions receiving federal funding for research. The Department of Heath, Education, and Welfare–NIH guidelines are published (Federal Register 41[131]: 27902–27943). Enforcement relies to a great extent on existing statutes and regulatory powers. Although restricting many categories of experiments, the guidelines do not go far enough for some people and are unnecessarily confining and bureaucratic to others. Industry is expected to comply voluntarily with the regulations, which disturbs many people who are concerned that the profit motive will override notions of safety in bringing products to market. The stringency of the guidelines applied to recombinant DNA research is later amended as experimental evidence accrues to support the safety of some types of experiments. Commercial release of recombinant organisms into the environment is regulated by the network of government agencies empowered by the numerous statutes covering biological and chemical products.

Robert Swanson (an investment broker) and Herbert Boyer (a molecular biologist involved in the first genetic engineering experiments) found Genentech, the first company based on genetic engineering technology, to produce medically important molecules. Thousands of other biotechnology companies will rise and fall through the years, but Genentech, the first and still the largest, will market some of the first genetically engineered biological products, mostly hormones or growth factors, produced by the new technology.

1977   A U.S. National Academy of Science meeting on recombinant DNA is held in Washington, D.C., March 7–9. Although intended as a forum on industrial applications of the new genetic engineering technology, major figures such as the Nobel Prize winner George Wald and his wife Ruth Hubbard, both committed foes of recombinant DNA technology, speak to the audience, turning the panel discussions into a debate and encouraging dissenters in the audience, including the activist Jeremy Rifkin, to take advantage of the media presence to support their cause. Rifkin argues that the nature of life itself

1977   is at stake and that it is necessary for the people to keep
(*cont.*) inhuman science in its place. Sixteen bills are introduced
           in Congress to regulate recombinant DNA research; none
           are ever passed into law.

           The first recombinant DNA molecule incorporating
           mammalian DNA is produced with genes for hemoglo-
           bin. The molecular basis for diseases with special impact
           on ethnic populations, sickle cell anemia (African de-
           scent) and beta-thalessemia (Mediterranean descent), is
           shown by DNA sequencing procedures developed the
           same year to be single amino acid changes in these vital
           oxygen-carrying proteins.

1978   The Nobel Prize in Medicine is awarded to Werner
           Arber, Daniel Nathans, and Hamilton Smith for the dis-
           covery and use of restriction enzymes for genetic engi-
           neering. A number of human gene products, including
           somatostatin and insulin, are produced by recombinant
           DNA techniques. The RAC is expanded to include mem-
           bers of the general public.

           The first baby conceived by mixing human sperm and a
           human egg outside of the body (in vitro fertilization, or
           IVF) is born in the United Kingdom. Nineteen years later,
           a sheep is "cloned" by transfer of an intact adult cell nu-
           cleus into an egg cell (ovum) with its nucleus removed.

1979   The NIH guidelines for recombinant methods of han-
           dling viral DNAs are relaxed with the accumulation of
           safety data and the development of impaired host organ-
           isms and engineered vectors. Cancer-causing genes are
           studied by transformation of cultured cells with DNA
           from malignant (cancerous) cells. Ironically, these are the
           types of experiments that in the summer of 1971 had set
           Robert Pollack and Paul Berg to contemplating the possi-
           ble ramifications of recombinant technology applied to
           cancer genes, leading to the Asilomar Conferences.

1980   The Nobel Prize in Chemistry is awarded to Paul Berg,
           Walter Gilbert, and Frederick Sanger for creation of the
           first recombinant DNA molecules and for DNA sequenc-

ing methods. U.S. molecular biologist Kary Mullis and others at the Cetus Corporation in Berkeley, California, invent the polymerase chain reaction (PCR) technique of replicating selected DNA sequences from a mixture of DNAs. This methodology revolutionizes molecular biology in the 1980s. The patent for the PCR process was later sold to Hoffman-LaRoche, Inc., in 1991 for $300 million. Biogen, another of the early genetic engineering firms, is founded by U.S. molecular biologists Walter Gilbert and Charles Weissman and begins producing interferon, a potent antiviral protein. Revised NIH Guidelines are published including voluntary compliance by non-NIH-funded institutions (Federal Register 45:[20]: 6724–6749, and revised Federal Register 45[227]:7 7384–77409).

1981   The first transgenic mammals are produced at Ohio University when foreign genes are transferred into mice. Golden carp produced by Chinese scientists are the first example of cloned fish. Sickle cell anemia becomes the first genetic illness diagnosed before birth at the gene level by restriction enzyme analysis of DNA. Congressman Al Gore begins a series of hearings on the relationship between academia and the growing commercialization of biomedical research. Such concerns are the harbinger of the university-industry web linking the many hundreds of new biotechnology companies with their academic founders and established industries. The fear that these interconnections will drastically influence the university research environment proves prophetic. The NIH publishes a final plan to assess formally the risks of recombinant DNA research (Federal Register 46[111]: 30772–30778).

1982   The pharmaceutical company Eli Lilly markets human insulin produced by recombinant DNA methods (Humulin). In addition to relieving a forecasted shortage of insulin for the increasing number of people with diabetes, allergic reactions to animal insulin are reduced by providing the human protein sequence. This hormone is the first of many biopharmaceuticals, small- to medium-sized proteins with potent biological effects on particular

1982   cell types, previously available only in minute quantities,
(*cont.*)   introduced into medical therapy.

A foreign gene cloned into tobacco plants is successfully
transmitted to progeny in common Mendelian fashion
just like indigenous genes. Besides showing the similar-
ity of genetically engineered changes to "normal" genetics, the transfer demonstrates the practicality of engineering plant genes that would soon become of major
economic and ecological interest. The following year,
U.S. patents are granted to companies for genetically engineered plants.

A human bladder cancer gene cloned into *E. coli* containing a single base pair change that resulted in a single
amino acid change in the protein is shown to be responsible for the cancer-causing activity of the protein when
expressed in mammalian cells. This observation suggests
that some cancers could be genetic diseases. AIDS is recognized as a syndrome, although the causative agent
will not be identified until the following year.

University of California scientist Stephen Lindow is the
first to ask permission to deliberately release genetically
modified organisms into the environment to prevent
frost damage on potatoes and strawberries, stimulating a
storm of controversy. The NIH approves his protocol the
following year.

1983   A chemical method for preparing synthetic genes is invented by chemist Marvin Caruthers at the University of
Colorado. Leroy Hood's automation of the procedure at
California Institute of Technology makes gene synthesis
a routine genetic engineering tool.

1984   Several moves are made to establish guidelines for controlling the impact of new medical technologies on humans. Activist Jeremy Rifkin authors a resolution signed
by 56 religious leaders and 8 scientists supporting somatic cell modification while opposing germ line (trans-
missible to subsequent generations) treatments as
human gene therapy. The National Organ Transplanta-

tion Act passes under the auspices of Senator Al Gore, prohibiting interstate commerce in organs or organ subparts for profit but allows transfers for medically relevant research purposes. The Warnock Report issued by the Parliamentary Committee of Inquiry in England recommends limiting research on human embryos.

1985   Congress authorizes the Biomedical Ethics Review Board to examine the ethical implications of genetic engineering of humans and to recommend policies and legislation to control potential abuses of the technology. The Biotechnology Science Coordinating Committee (BSCC) is established under Senator Al Gore as part of the U.S. Government Office of Science and Technology Policy.

Federal courts rule that private companies do not need NIH approval for field testing of genetically engineered organisms. EPA approval has been required since 1984.

MVP, a genetically engineered biopesticide, a protein toxic to certain insects, developed by Mycogen, becomes the first such product approved by the EPA.

1986   Amid the controversy over relationships between universities and industry, CRADAs (Cooperative Research and Development Agreements) are instituted through the Technology Transfer Act of 1986 to facilitate technology transfer from government (NIH) to industry. Although this enables high-ranking government scientists to tap more private funds for research, the potential conflict of interest erodes public confidence in personal and government impartiality. The first field trials of genetically engineered plants resistant to insects, viruses, and bacteria are undertaken. In a statement he has regretted since, Walter Gilbert calls the complete human genome sequence the "holy grail" of biology.

1988   James D. Watson, then director of Cold Spring Harbor Laboratories where much of the new recombinant DNA technology had its start, is appointed head of the National Center for Human Genome Research. It was felt that a scientist with the stature of a discoverer of the double helical

1988   structure of DNA was needed to galvanize the massive ef-
(*cont.*)   fort required to determine the complete DNA sequence of
        the human genetic complement, the genome. The first
        patent on a living transgenic animal, the Harvard Onco-
        Mouse (US 4,736,866), is issued to Phil Leder and Harvard
        University. This mouse is an animal model for antitumor
        therapies that had been engineered to be highly suscepti-
        ble to certain types of cancer. The OncoMouse patent was
        refused in Europe.

1990   Although most critics worry about private industry com-
        mercializing genetic engineering, the first gene patents
        are actually obtained by the U.S. government's (NIH)
        Craig Venter, who patents thousands of genes based on
        short pieces of DNA sequence with functions that can be
        predicted. Venter soon leaves NIH to start a biotechnol-
        ogy company outside of Washington, D.C., in Gaithers-
        burg, Maryland, based on these "expressed sequence"
        tags (ESTs). This is the year the Human Genome Project
        begins, a 15-year project to sequence the entire human
        genome. The Office of Technology Assessment reveals
        that 13% of Fortune 500 companies are using or had used
        some form of genetic screening in hiring decisions.

        A young girl, Ashanthi DeSilva, receives the first cellular
        gene therapy using her own white blood cells genetically
        modified to regain their immune function resulting from
        an adenosine deaminase (ADA) enzyme deficiency.
        Modest improvement is seen, but, more important, no ill
        effects are noted. Numerous other gene therapy proto-
        cols soon follow, directed against intractable diseases
        such as cancer, Duchenne's muscular dystrophy, and
        cystic fibrosis as clinicians struggle to learn if the new
        technology will live up to its potential.

        In the United Kingdom, the Human Fertilization and
        Embryology Act of 1990 bans human cloning.

1991   The Task Force on Genetics and Insurance is established
        at the National Center for Genome Research (NIH/De-
        partment of Energy) to address issues raised by genetic
        testing. This is part of the comprehensive plan to control

consequences of the explosion of knowledge from the Genome Project.

1992   The Recombinant DNA Advisory Committee is disbanded and its functions taken over by various government agencies. The RAC served through the turbulent years of assessment of the safety of the new recombinant DNA technology while the regulatory aspects were being developed.

Wisconsin is the first state to forbid discrimination in employment or insurance based on an individual's genetic background. Other states steadily follow suit. Loopholes remaining for company-financed insurance plans eventually lead to introduction of protective federal legislation in 1997.

1993   Despite heavy controversy, technology transfer continues unabated with 200 CRADAs issued. The genetically engineered delayed-ripening Flavr Savr/MacGregor tomatoes are marketed by Calgene.

1994   Fifty-nine gene therapy protocols involving 150 patients are approved.

1995   The first complete nucleotide sequence of a bacterium (*Hemophilus influenzae*) is published. One hundred and six clinical protocols involving gene therapy had been approved and 597 patients had undergone experimental gene transfer since the first gene therapy experiment in September 1990. NIH spends $200 million per year on gene therapy research. Biotechnology companies are thought to be spending as much again or more. The NIH RAC concludes after a study of gene therapy that, despite great promise and expectations, clinical efficacy had not been demonstrated definitively in any gene therapy protocol.

1997   The cloning of a sheep from a somatic cell nucleus in Scotland raises a storm of concern over human cloning, previously ignored because it was so far from realization. Cloned rhesus monkeys are reported. Suddenly human

1997   cloning appears imminent, although formidable techni-
(*cont.*) cal obstacles remain. Federal legislators rush to intro-
        duce a number of bills prohibiting human cloning. Pro-
        tection against genetic discrimination is improved by the
        introduction of several Genetic Confidentiality and
        Nondiscrimination bills.

        Although the public is willing to accept genetically engi-
        neered medical products, it is another matter entirely
        when it comes to what it eats. Reaction against geneti-
        cally engineered foods in the United States results in the
        Bovine Growth Hormone Milk Act regulating labeling of
        milk products produced with the aid of the synthetic
        hormone. The European Union bans milk or meat prod-
        ucts produced with synthetic growth hormone and
        maintains a similar stance on food products from genet-
        ically engineered plants.

        Flavr Savr tomatoes are removed from the market for
        "improvements." They never return.

1998   Human cloning remains a topical issue. The mouse, an
        important laboratory animal, is cloned by nuclear trans-
        plantation, a process that is expected to speed up certain
        kinds of research. Reaction to the pending blanket human
        anticloning legislation by the clinical and basic science
        community as well as supporters of medical research
        calls for further study of the proposals. The intention is to
        avoid outlawing vital scientific and medical procedures
        needed to develop new therapies while maintaining a
        stance against cloning of human beings.

        Human embryonic stem cell lines are established by two
        groups, one at the Geron Corporation (Menlo Park, Cal-
        ifornia) and in the laboratory of James Thomson at the
        University of Wisconsin. The controversy begins.

        The first complete animal genome, that of the worm
        Caenorhabditis elegans, is sequenced, and a rough map of
        the human genome is published showing the location of
        nearly 30,000 genes.

2000   The first complete genome map of a plant, *Arabidopsis thaliana,* is published, a task made difficult by the large size of plant genomes. Genetically engineered crops are grown on 108.9 million acres in 13 countries.

On June 25, 2000, the International Human Genome Project and the private Celera Genome Corporation jointly announce completing the initial sequencing of the human genome.

In August, the Clinton administration publishes federal guidelines for funding of stem cell research.

2001   First complete map of a food plant, rice, is completed, and the genomes of several agriculturally important bacteria, a nitrogen-assimilator and a plant pathogen, are published. On August 9, 2001, President George W. Bush announces his decision to allow federal funding of the use of existing pluripotent (not totipotent) stem cell lines derived from human embryos prior to that date.

2002   Many other genome sequences are reported including that of the malaria parasite and the mosquito that carries it, a pathogen of rice, and a draft version of the full human genome. The mouse genome project publishes a draft sequence.

A regulatory role is discovered for previously small RNA sequences in controlling cellular functions that had been puzzling for many years.

Genetically engineered crops are grown on 145 million acres in 16 countries.

2003   Disease mapping studies using the human genome sequence identify a gene for vulnerability to depression, and linkages are sought for schizophrenia and bipolar disorder, two other psychiatric disorders. Genetically engineered crops are grown on 167.2 million acres in 18 countries. The EPA approves the first transgenic rootworm-resistant corn expected to save farmers more than

2003
(*cont.*)
$1 billion a year in pesticides and crop losses. A genetically engineered coffee plant that produces caffeine-free beans is developed in Japan.

Dolly, a sheep, the first cloned mammal (1997) is euthanized after developing a progressive lung disease and other signs of premature aging.

2004
The first human stem cell line produced with the aid of somatic nuclear transfer in Korea.

The chicken genome sequence is published.

Agriculture biotechnology receives a boost from the results of studies carried out by influential groups. The United Nations Food and Agriculture Organization endorses biotech crops as a complement to traditional farming. The National Academy of Science Institute of Medicine finds that biotech crops do not pose any more danger to consumers than traditional crops. Genetically engineered crops are grown on 200.5 million acres worldwide.

5

Biographical Sketches

William French Anderson (1936–)

W. French Anderson grew up in Tulsa, Oklahoma, where he was born on December 31, 1936. He decided that he wanted to work on curing genetic diseases after finishing an undergraduate course at Harvard on DNA and genetics taught by James D. Watson, the Nobel Laureate. After graduating with an A.B. in 1958, he studied with Francis Crick at Cambridge, receiving an M.A. in 1960. Then it was back to Harvard for an M.D. in 1963 followed by a pediatric residency at Boston Children's Hospital. His research career began in 1965 at NIH, where he studied hemoglobin synthesis and the inherited blood disorders thalassemia and sickle cell disease, which result from mutant hemoglobin molecules. Anderson pioneered the use of iron-binding drugs in iron overload conditions. In 1977 he became director of the Molecular Heredity Section at the NIH. Searching for practical ways to introduce genes into cells after removing the harmful genes, Anderson began to test retroviruses, a family of RNA viruses that included members that normally cause tumors and AIDS as potential therapeutic vectors. The hemoglobin disorders proved too complex for beginning gene therapy in humans, so Anderson selected the newly discovered adenosine deaminase gene for replacement in patients immunocompromised by an ADA gene deficiency, a rare genetic disorder. Following successful gene marking tests with the engineered viral vector system in terminal cancer patients in 1988, the first true gene therapy treatment was initiated by Drs. Anderson, R. Michael Blaese, and

Kenneth W. Culver on September 14, 1990. The ADA gene in a retroviral vector was inserted into some of a nine-year-old girl's own white blood cells, which were then returned to her bloodstream. Blood ADA levels rose and the her immune system partially recovered—a success!

In 1991, Anderson founded the private company Genetic Therapy to commercialize work that was being done in federal government laboratories and to provide vectors for gene therapy to hospitals and universities. He moved to the University of Southern California in 1992.

Paul Berg (1926–)

Paul Berg, currently at Stanford University where he has been on the faculty at the medical center since 1959, was born in New York City on June 30, 1926. He received his B.S. at Penn State in 1948 and his Ph.D. at (Case) Western Reserve University in 1952. Following postdoctoral studies at Washington University in St. Louis, he joined the faculty there in 1955 where he remained until he moved to Stanford as an associate professor in 1959. At that research mecca, Berg found himself in the midst of scientific developments that would lead to recombinant DNA technology. He was recognized early on for his seminal contributions to the determination of the mechanism by which proteins are coded by DNA. Before age 40, Berg was elected to the prestigious U.S. National Academy of Sciences, just one of a long series of awards that have acknowledged his contributions to science and society.

It was Berg's concern about the safety of certain kinds of DNA manipulation that he was contemplating, particularly tumor virus DNA, that lead to the historic "Berg letter" to *Science* and the call for a moratorium on certain recombinant DNA research until safety issues were addressed. He was a key organizer of the Asilomar Conference in 1975, which lead to the National Institutes of Health guidelines on recombinant DNA research issued in 1976, a historic instance of responsible self-regulation in scientific research. Paul Berg was awarded the Nobel Prize in Chemistry in 1980 for his work on DNA. This recognition for his scientific accomplishments, among others, was followed by acknowledgment of his social contributions with the Scientific Freedom and Responsibility Award, the National Medal of Science, and the National Library of Medicine Medal. His studies continued to involve recombinant DNA technology through the 1980s. In 1985, Berg was appointed to direct the Beckman Center for Molecular and Genetic Medicine at Stanford. Continuing his socially responsible ethic, in 1991 he was named as the head of the Human Genome Project Scientific Advisory Committee of the National Institutes of Health. In this role, he has been considering the ethical and practical issues raised by the new technology and the impact of genetic technologies on society.

Herbert Wayne Boyer (1936–)

Herbert Boyer was born in Pittsburgh, Pennsylvania, on July 10, 1936, and attended tiny St. Vincent College in nearby Latrobe, Pennsylvania, where he received his A.B. in biology and chemistry in 1958. The former high school football lineman went on to the University of Pittsburgh where he earned his M.S. in 1960 and finally his Ph.D. in bacteriology in 1963. After 3 years of postdoctoral experience at Yale University, he moved to the Department of Microbiology at the University of California at San Francisco in 1966 where he became embroiled in the early events of the recombinant DNA revolution. His laboratory was investigating the restriction enzymes of *Escherichia coli,* discovering that Eco RI (*E. coli* restriction endonuclease number I) cut DNA into reproducible-sized pieces, producing "sticky ends" that tended to adhere to other pieces of DNA treated with Eco RI. Boyer began collaborating with Dr. Stanley Cohen at nearby Stanford, who had been studying plasmids, circular pieces of DNA carrying genes for such properties as antibiotic resistance in bacteria. The laboratories of Boyer and Cohen were soon able to recombine segments of DNA in a desired order to include genes for proteins and to put the engineered plasmids back into bacteria. These altered bacteria, when properly grown, would then make that protein under the direction of the plasmid DNA regardless of the origin of the protein because the genetic code for the various amino acids was universal. This process was to provide the basis for the biotechnology industry.

In 1976, Boyer cofounded Genentech along with venture capitalist Robert Swanson on the strength of the new technology and served as vice president of the company until 1990 when he shifted to the board of directors, where he remains at present. Genentech produced the first biopharmaceutical drug in 1978, the peptide hormone human insulin, which was subsequently licensed to the pharmaceutical company Eli Lilly. Genentech was the first biotechnology company to produce its own product when it launched human growth hormone in 1985. Boyer retained his appointment at the University of California at San Francisco, becoming a full professor in the Genetics Division of the Department of Biochemistry and Biophysics in 1976; he is now a professor emeritus there. He has been the recipient of numerous academic and industrial honors and prizes for his pioneering applications of genetic engineering and is still on the board of directors at Genentech. Boyer is a member of the National Academy of Sciences, a fellow of the American Academy of Arts and Sciences, and a recipient of the Albert Lasker Basic Medical Research Award, considered by many to be the highest scientific award short of the Nobel Prize.

Ananda Mohan Chakrabarty (1938–)

Ananda Chakrabarty was born in Sainthia, India, on April 4, 1938. He earned his B.Sc. from St. Xavier's College in Calcutta, India, in 1958, and both his M.Sc. (1960) and his Ph.D. in biochemistry from Calcutta University in India (1965). After a year as a scientific officer at Calcutta University, Chakrabarty moved to the United States as a research associate at the University of Illinois at Urbana, where he worked from 1965 to 1971. He then made the transition to industry, joining the General Electric Company in Schenectady, New York, as a staff microbiologist dealing with environmental pollutants. He recognized the potential for bioremediation of contaminants through the metabolic activity of microbes and created strains of *Pseudomonas* bearing plasmids coding for enzyme activities capable of converting toxic oil spills to harmless by-products. A landmark patent case claiming specific genetically engineered organisms to clean up oil spills was filed in 1971 by General Electric and Chakrabarty. The original application was rejected but was finally upheld by the Court of Customs and Patent Appeals in 1980, the year after Chakrabarty left General Electric to join the faculty in the Department of Microbiology at the University of Illinois Medical Center in Chicago, where he teaches and does research today. For his work with the oil-eating bacteria, Chakrabarty was recognized as Industrial Scientist of the Year in 1975 and later was honored with the Public Affairs Award from the American Chemical Society (1984) and the Pasteur Award (1991).

The U.S. Supreme Court turned back yet another challenge to the patent in 1990. This key patent decision removed the distinction between the patenting of animate and inanimate inventions. The possibility of protecting the commercial potential of genetically engineered organisms is a cornerstone of the biotechnology industry.

John William Coleman (1929–)

John Coleman was born in New York, New York, on December 30, 1929. As a black man in the 1940s, his educational choices were initially limited. He received his bachelor's degree from Howard University in 1950. He served as a physicist for the U.S. National Bureau of Standards from 1951 to 1953 and then as an instructor in physics at Howard University from 1957 to 1958. Coleman continued to advance his studies during this time, obtaining his master's degree from the University of Illinois in 1957. In 1958, he was hired as an engineer by the RCA Company, where he focused his research on the physics of electrons. During his tenure at RCA, he earned his Ph.D. in biophysics at the University of Pennsylvania, receiving his degree in 1963. Although the electron microscope was invented in 1934 by Max Knoll and Ernst Roska, it remained a physics curiosity until 1940 when RCA demonstrated a crude commercial version. Coleman was involved in the development of the electron microscope at RCA as it was transformed from the behemoth 2.5-ton instrument (1949) suitable for materials science study (metallurgy, ceramic surfaces) into a high-resolution device useable for properly treated biological samples.

Francis Sellers Collins (1950–)

Francis Collins was born in Staunton, Virginia, on April 14, 1950. He was educated at the University of Virginia in Charlottesville, receiving his B.S. in 1970, his M.S. in 1972, and his Ph.D. in 1974 from Yale University. Collins earned his M.D. from the University of North Carolina at Chapel Hill in 1977 and then served his internship and residency at the hospital there through 1981. He continued his medical education as a fellow in medical genetics and pediatrics from 1981 to 1984 at the Yale University School of Medicine. He moved to the University of Michigan in Ann Arbor in 1984 as a faculty member in the Departments of Internal Medicine and Medical Genetics, serving as chief of those departments from 1987 to 1991 as well as a Howard Hughes Investigator at the University of Michigan from 1987 to 1993. In 1993, he replaced the flamboyant and controversial James D. Watson as the director of the National Center for Human Genome Research at the National Institutes of Health in Bethesda, Maryland.

Numerous honors and awards have been bestowed on Dr. Collins, and he is on the editorial boards of several prestigious scientific journals dealing with human genetics, molecular biology, and genetic engineering. He is a member of the Institute of Medicine of the U.S. National Academy of Sciences.

While at the University of Michigan, Dr. Collins became known for cloning the most prevalent form of the defective gene responsible for 80% of the cystic fibrosis (CF) cases, seen at that time as a scientific coup and an example of the power of genetic techniques for elucidating the cause of diseases. It also raised the possibility of genetic therapy for this commonly inherited (1:2,000 Caucasian births) disease. Despite much effort, however, successful clinical treatment by gene therapy for this condition remains to be demonstrated. Besides his work on CF, Dr. Collins is a key figure in the massive Human Genome Project, maintaining a constant presence in the public eye and balancing the swirling issues of ethics, morality, and privacy while coordinating one of the most ambitious human endeavors ever: sequencing the human genome by the year 2006. Why would anyone take on this tremendous responsibility? Collins accepted a pay cut to direct the Genome Project, saying, "I feel I've been preparing for this job my whole life."

Francis Harry Compton Crick (1916–2004)

Francis Crick was born in Northhampton, England, on June 8, 1916. As a boy, he was fascinated with science but was overwhelmed by the fear that by the time he grew up, everything would have been discovered. He trained in physics, receiving his B.Sc. from the University College of London in 1937. During and after World War II (1940 to 1947), he served the British Admiralty as a scientist. He was a medical research council student at the

Strangeways Laboratory at Cambridge from 1947 to 1949 and then on staff at the Molecular Biology Laboratory at Cambridge from 1949 to 1977. He began his famous collaboration with James D. Watson on the architecture of DNA, contributing his expertise in X-ray crystallography, which resulted in publication of the double helical model for DNA structure in 1953. Crick received his Ph.D. from Cambridge in 1954 and continued his work on the chemical basis for DNA structure. In 1955, he postulated that proteins were produced by adaptor molecules from the DNA code. By 1958, there was enough information to support the hypothesis that biological information flowed from DNA through RNA to protein structure, a hypothesis that became known as the central dogma of molecular biology. The eventual discovery of viruses in which the genetic repository was RNA (hence the name *retroviruses* for some) required qualification of the hypothesis, but the concept still generally holds. The 1962 Nobel Prize in Physiology and Medicine was awarded jointly to Crick, Watson, and Wilkins for their work in determining the structure of DNA and the catalytic impact on molecular biology and biochemistry. Many scientific honors followed over the years, including election to the American, German, French, and Indian Academies of Science and selection as a fellow of the Royal Academy of Science.

In 1977, Crick moved to the Salk Institute for Biological Studies in La Jolla, California, and began to engage his interests in the neurosciences and complex behavior such as consciousness as set forth in his *The Astonishing Hypothesis* (Scribner, 1994). Crick's quiet style of doing science never earned him the antipathy that his more outspoken collaborator, James Watson, seemed to relish. He lost a several-year battle with colon cancer on July 28, 2004.

Rosalind Franklin (1920–1958)

Rosalind Franklin was born on July 25, 1920, in England. She graduated from the Cambridge University undergraduate program and began graduate work in physical chemistry. While still a graduate student, Franklin made her contribution to the World War II effort as an assistant research officer of the British Coal Utilization Research Association. From 1942 to 1946, she did pioneering work on coal microstructure. In 1945, she received her Ph.D. In February 1947, Franklin went to the Laboratoire Central des Services Chimiques de l'Etat in Paris, France, to learn X-ray diffraction techniques to further her microstructure work. After several productive years, she was eager to tackle biological structures and took up John Randall's offer to set up an X-ray diffraction facility in Maurice Wilkins's laboratory at King's College, England, in 1951. She succeeded in obtaining high-quality diffraction patterns of oriented B-form DNA fibers, clearly recognizing the helical organization of the fiber and postulating a multichain helix with the phosphate backbone on the outside and the nucleic acid bases on the inside. This information was communicated at a colloquium in November 1951 at which James Watson was present. From this information, from Maurice Wilkins's communications, and from official King's College laboratory reports, combined with their own data and intuition, Watson and Crick built the models that culminated in the publication of the helical paradigm of DNA structure.

Despite not receiving full credit for providing the experimental evidence of the structure of DNA, Franklin continued her work and published a structure for the A-form of DNA. In 1953, she left King's College for Birkbeck College in London to work with J. D. Bernel's group where she produced one of the first X-ray structures of the tobacco mosaic virus, ironically comprising protein subunits arranged in a hollow helix around a DNA core. Just a few years later, on April 16, 1958, Rosalind Franklin died of cancer.

The Nobel Prize for the structure of DNA was shared by Watson, Crick, and Wilkins in 1962. Although the Nobel Prize is only awarded to living persons, many people attribute the apparent snub to the fact that Rosalind Franklin was a woman.

Albert Gore, Jr. (1948–)

Albert Gore, vice president of the United States from 1992 to 2000, is a politician who has had significant impact on the recombinant DNA debate through his involvement in ecological issues while he was in Congress. Born in Washington, D.C., on March 31, 1948, he obtained his A.B. from Harvard University in 1969 and then served a stint in Vietnam with the U.S. Army (1969–1971). Returning from the service, he did postgraduate work at the Graduate School of Religion at Vanderbilt University (1971–1972) and then at the law school there from 1974 to 1976.

During this period, he was an investigative reporter and editorialist for *The Tennessean* (1971–1976) where he honed his writing skills. Gore entered public service in the U.S. House of Representatives (1977–1985) and then as a senator (1985–1993). As a congressman, he conducted a series of hearings on the connection between academia and industry in the commercialization of biomedical research, and as a senator he was involved with the National Organ Transplantation Act to regulate the commercialization of human organs for transplants. In Congress he established himself as an expert on the environment and on nuclear arms control as well as bioethics. Gore is the author of *Earth in the Balance: Ecology and the Human Spirit* (1992).

Leroy E. Hood (1938–)

Lee Hood was born in Missoula, Montana, on October 10, 1938. He received a B.S. from the California Institute of Technology in Pasadena, California, in 1960 and proceeded to the Johns Hopkins University School of Medicine in Baltimore, Maryland, where he earned his M.D. in 1964. He then returned to Caltech where he obtained his Ph.D. in biochemistry in 1968 and cultivated his interest in molecular immunology. From 1967 to 1970, he was a fellow at the National Cancer Institute studying immunology. Still drawn to California, Lee returned once more to Caltech, this time as a faculty member where he rose to the rank of full professor by 1977. In 1989, he was appointed director of the National Science Foundation Center for Molecular Biotechnology. In 1992, he accepted the chairmanship of the new Molecular Biotechnology Department at Caltech.

Hood's interest in the sequence basis for the polymorphic immune system made him impatient with the available technology for protein and nucleic acid sequence analysis. He began designing and building instrumentation that could automatically and rapidly determine the nucleic acid sequences of DNA and the amino acid sequences of proteins from very small amounts of sample. He also built machines that could chemically make synthetic nucleic acids (DNA and RNA) and polypeptides (proteins) from their component monomer units. He also commercialized these instruments so that they were available to the scientific community where they made possible the rapid progress in sequencing the new genes being discovered daily. The automated nucleic acid sequencing technology has been at the core of the Human Genome Project to sequence the human gene complement, without which the project would not be possible.

Dr. Hood is a member of the U.S. National Academy of Sciences. He was awarded the Louis Pasteur Award for Medical Innovation and the Albert Lasker Basic Medical Research in 1987 for his studies of immunological diversity. The technological advances spawned by the instrumentation and techniques developed in Hood's laboratory have earned him numerous other awards and honors. He is becoming increasingly involved in the analysis of the genetic information that is accumulating from the genome sequencing projects from multiple organisms, realizing that effectively utilizing this information to solve important medical problems is as big a challenge as acquiring the information in the first place.

Edward Moore Kennedy (1932–)

Edward (Ted) Kennedy was born in Boston, Massachusetts, on February 22, 1932, the son of Joseph and Rose Kennedy. He was born into a prominent family steeped in politics and with a tradition of public service. The older brother of Robert (U.S. attorney general) and John F. Kennedy (U.S. president), both victims of assassin's bullets, obtained his A.B. from Harvard University in 1956. He pursued training in the law with postgraduate study at the International Law School at The Hague, the Netherlands, in 1958, obtaining his L.L.B. in 1959 from the University of Virginia. Passing the Massachusetts bar exam in 1959, he was the assistant district attorney for Suffolk County from 1961 to 1962. He was elected U.S. senator from Massachusetts in 1962 and has remained an outspoken liberal voice in that body on social issues since that time. During his tenure in the Senate, he has served on the Judiciary Committee (1979–1981), the Armed Services Committee, and the Democratic Steering and Organization Committee. He has been particularly involved with health issues and was a prominent figure in Congress as chairman (1971–1980) of the Subcommittee on Health of the Labor and Human Resources Committee during the early attempts to legislate regulation of the use of recombinant DNA technology. He is also a member of the Biomedical Ethics Board. His publications include *Decisions for a*

Decade (1968), *In Critical Condition: The Crisis in America's Healthcare* (1972), *Our Day and Generation* (1979, with Mark O. Hatfield), and *Freeze: How You Can Help Prevent Nuclear War* (1979).

Michael Martin (1956–)

Michael Martin was born in 1956 in the rural Alabama town of Orrville, growing up on a small cotton and livestock farm. Martin's encounter with the southern corn leaf blight of 1970, which destroyed his hopes of winning a Future Farmers of America award, left him searching for an explanation of why some corn varieties survived and others were wiped out. A county extension agent explained the basis of genetic resistance to the black teenager, setting him onto a career in crop improvement. Martin earned a B.S. in agronomy from Alabama A&M University and went on to Iowa State for an M.S. and eventually his Ph.D. (1982) in plant breeding and cytogenetics. One of his former professors hired the newly graduated Martin as a research center manager for the Garst Seed Company, a leader in hybrid corn development. Garst merged with Zeneca Seed of Britain in 1990 and continued to grow. Martin led Garst into the first commercial introduction of an herbicide-resistant as well as multigenic disease-, insect-, and climate-resistant corn. By 1996, he was supervising a staff of 247 and an annual budget of $18.5 million. The formation of Advanta Seeds by Zeneca and the Dutch Royal VanderHave Group further expanded his horizons and responsibilities. As the world's largest provider of canola and sunflower seeds, and ranking in the top five worldwide in corn seeds, Advanta Seeds has projected annual sales in 1997 of more than $500 million. Martin is responsible for developing and implementing a research, sales, and distribution plan in his specialty of dent and tropical corn varieties.

Barbara McClintock (1902–1992)

Barbara McClintock was born June 16, 1902, in Hartford, Connecticut. After graduating from high school at age 16, she worked at an employment agency until she entered Cornell University's Agriculture College. She decided to study genetics, receiving her B.S. in 1923. She was fascinated by the chromosomes of corn plants; maize was the genetic system studied at Cornell at the time rather than the fruit fly, *Drosophila melanogaster,* which was in vogue at other universities. Careful microscopic study of maize development allowed McClintock to see clearly the chromosomes divide and rearrange. The characterization of this system became her Ph.D. dissertation project, which she earned through the Department of Botany in 1927. She remained as an instructor until 1931 when she and Harriet Creighton published work that connected the physical interchange of maize chromosome material with the exchange of genetic information, just weeks ahead of the publication of similar observations with *Drosophila.* Very much a loner and uncomfortable with the norms of scientific behavior, McClintock held various positions and did research at Cornell, the University of Missouri, and CalTech between 1931 and 1940. Her reputation as a cytogeneticist was impeccable. In 1941, she was invited to join the staff of the Carnegie Institute of Washington (D.C.) Department of Genetics at the Cold Spring Harbor Laboratories on Long Island, New York, where she remained for the rest of her life.

McClintock was elected to the U.S. National Academy of Sciences in 1944, the third woman so honored. The subsequent year, she was elected president of the prestigious Genetics Society of America, the same year that genes were demonstrated to be made of DNA. In 1951, McClintock tried to explain her observations on the transmission of certain maize characteristics by postulating that DNA could be rearranged on the scale of genes, but she failed to gain the general acceptance of the scientific community. A number of years later, after molecular techniques demonstrated similar events with bacteria and viruses, McClintock's data on evolutionarily more advanced organisms obtained with 19th-century equipment, a regular microscope, crossbreeding experiments, and careful observation finally gained the recognition it deserved. DNA can rearrange—"jumping genes" are a biological reality. Similar DNA rearrangements within a gene of the white blood cells of the immune system are crucial for the construction of antibodies that protect against foreign invaders. McClintock was honored for her contributions, receiving several awards including the National Medal of Science in 1970 and finally winning the Nobel Prize in Physiology and Medicine in 1983. With all of the attention so richly deserved, Barbara McClintock remained quietly aloof, working in her laboratory and cornfields until she died on September 2, 1992.

Louis Pasteur (1822–1895)

Louis Pasteur was born on December 27, 1822, at Dole in the Jura region of France. An unremarkable student until near the end of his secondary school career, Pasteur obtained his bachelor's degree in 1840, after a series of fits and starts, from the Collège Royal de Besançon. He eventually entered the École Normale Supérieure in Paris, receiving his doctorate in physics and chemistry in 1847. He continued to work at the École Normale while awaiting a faculty position, becoming involved in studies on the optical activity of molecules. This work became the starting point for his career-long involvement in seemingly diverse and unrelated research topics. In 1849, he became professor of chemistry at Strasbourg and then professor and dean of the Faculty of Science at Lille in the industrial region of France in 1854.

Associated with an institution closely allied with manufacturing interests, Pasteur found an outlet for his highly practical inclinations in Lille. His investigations into the biological and chemical origins of optical activity and development of techniques for separating the components led Pasteur into the examination of fermentation reactions of various types. He conducted a series of experiments that led to elimination of spontaneous generation in 1860 as an explanation of spoilage processes plaguing the wine and vinegar industries. The heat treatment method that bears his name, pasteurization, proved effective for several industrial applications, and he eventually patented the procedure in 1865. In 1857, he moved back to the École Normale as director of scientific studies. He was asked by the French silkworm industry to find a solution to a mysterious disease that was ruining silk production. Through long and tedious experimentation, Pasteur was able to demonstrate two simultaneously occurring microbial diseases and to work out procedures to eliminate further outbreaks. During this time, the germ theory of disease was receiving experimental support from Robert Koch, the father of bacteriology, and others who, along with Pasteur, developed methods of growing pure cultures of disease-causing organisms. Joseph Lister introduced antiseptic surgery in which instruments, hands, and surroundings were chemically sterilized with carbolic acid in the 1860s. These precautions were immediately effective in reducing hospital patient mortality.

Pasteur began studies on anthrax, a deadly bacterial disease, after hiring assistants who could perform the necessary biological experiments, which he, partially paralyzed by a stroke in 1868 and as an antivivisectionist, could not. He became one of the pioneers of immunity and prophylaxis, following up on the successful late-18th-century smallpox vaccinations by Edward Jenner. A successful, highly publicized field trial of an attenuated anthrax treatment on susceptible farm animals in 1881 led Pasteur to extend his work to fowl cholera, hog cholera, and rabies. The July 6, 1885, treatment of nine-year-old Joseph Meister who had been bitten by a rabid dog with attenuated virus-infected spinal cord extracts prevented the inevitably fatal progression of the disease and solidified Pasteur's fame.

In 1888, Pasteur became director of a new institute bearing his name, dedicated to the treatment of rabies and the development of more effective vaccines. Increasingly incapacitated by illness, Pasteur continued directing research at the institute. He died near Paris on September 28, 1895, and was honored with a state funeral at the Cathédrale de Notre Dame. Pasteur was the recipient of numerous awards and much honorary recognition, including election to the Royal Society, to the Académie de Medécine, and to the Académie Française. His contributions to both the experimental demonstration of the germ theory of disease and to the practical development of treatments for microbial diseases laid the basis for modern immunoprophylaxis and therapy.

Jeremy Rifkin (1945–)

Growing up in the late 1940s to early 1950s on the south side of Chicago, Rifkin gave little hint of the political and social activist he would become. The son of a plastic bag manufacturer and a charity worker, he developed his antiwar and civil rights sentiments at the University of Pennsylvania's Wharton Business School of Finance. He feels that growing up in the Vietnam War protest era steered him to his activist profession. Rifkin served as an organizer of the 1968 March on the Pentagon and in 1969 founded the Citizen's Commission to focus on alleged U.S. war crimes in Vietnam. In 1971, he founded the People's Bicentennial Commission as a countercultural alternative to official government plans for the U.S. bicentennial celebration.

By the time the recombinant DNA controversy came on the scene, Rifkin was well entrenched as a foe of the establishment. He founded the nonprofit Foundation on Economic Trends, based in Washington, D.C., to influence government policy on a spectrum of economic, environmental, scientific, and technological issues. Although lacking formal training in the sciences, beginning in the late 1970s, he became a strident voice in opposing genetically engineered crops, patenting of genes, genetic engineering of animals and animal breeding practices, biological weapons, and the sale of recombinantly modified foods. Almost exactly 20 years predating the 1997 announcement of the cloning of the sheep Dolly by nuclear transfer, Rifkin and a group of protesters disrupted a National Academy of Sciences meeting on applications of recombinant DNA technology chanting, "We will not be cloned!" He apparently feels that biotechnology would allow scientists to "play God" and could lead to eugenic programs. Through his foundation he has filed lawsuits against various government agencies, including the Department of Agriculture on a number of fronts challenging the genetic engineering and animal breeding practice on the basis of animal rights, and various proposed releases of genetically modified organisms such as engineered frost-preventing bacteria (Frostban) and crude-oil-eating bacteria for oil spills on the basis of inadequate environmental impact assessment.

Although Rifkin moved on to other concerns including global economics, workers' rights issues, the impact of information technologies on the workplace, and global warming, his foundation in the 1990s began rallying grassroots opposition to the use of genetically engineered bovine growth hormone to enhance milk and meat production in cattle. In May 1995, he orchestrated the issuance of a statement from 180 religious leaders from some 80 religious groups including United Methodist, Southern Baptist, Jewish, and Muslim organizations, calling for a moratorium on patenting of genetically engineered animals and human genes, cells, tissues, organs, and embryos. In April 1997, the foundation, in concert with others, organized a global protest to oppose genetically engineered foods, cloning, and patenting of genes and life forms. Despite his fervent opposition to many aspects of genetic engineering, Rifkin maintains that he has never opposed biotechnology for genetic screening, applying genetic knowledge to preventative medicine, or for the production of pharmaceuticals.

By enlisting strong emotional opposition to genetic engineering and other issues, Rifkin has incurred the enmity of government officials, scientists, and industry executives to whom he is known as the "abominable no man." He responds by saying that critical perspective is needed in commercialization of genetic technologies, pointing to the lack of such a public debate on the nuclear and chemical technologies that led to the Three Mile Island, Chernobyl, and Bhopal disasters.

Rifkin is president of the Foundation on Economic Trends, president of the Greenhouse Crisis Foundation, and head of the Beyond Beef Coalition. He is the author of 13 books, including *Algeny* (Viking Press, 1983), which expressed a naturalistic or vitalistic view of nature and caused a stir by questioning the objectivity and validity of Darwinian evolution using creationism arguments. Rifkin returned at the close of the 1990s to stir the pot on genetic engineering as the author of *The Biotech Century: Harnessing the Gene and Remaking the World* (Putnam, 1998), on the effects of globalization on cultural identity in *The Age of Access* (Tarcher, 2001), on fossil fuels in *The Hydrogen Economy* (Tarcher, 2002), and on the lapsing of the American Dream and its replacement by the European Union in *The European Dream* (Tarcher, 2004)

Marian Lucy Rivas (1943–)

Marian Rivas was born on May 6, 1943, in New York City. After receiving her B.S. from Marian College in 1964, she continued her studies at Indiana University where she earned first an M.S. (1967) and then a Ph.D. (1969) in the then-new field of medical genetics. She extended her training with a fellowship at Johns Hopkins University from 1969 to 1971. Her first faculty position came at the Douglas College of Rutgers University in New Jersey where she was an assistant professor from 1971 to 1975, just as the recombinant DNA revolution was beginning to sweep biology. The medical applications of the new technology and their consequences were to become apparent just a few years later. In 1975, Rivas moved to the Hemophilia Center of the Oregon Health Sciences University where she has been a full professor since 1982. In 1978, she also became an associate scientist at the Neurologic Science Institute of the Good Samaritan Hospital. As a medical geneticist, Rivas has served on several committees on genetics at the National Institutes of Health. Her career has spanned human gene mapping and investigation of the genetic aspects of epilepsy, and she has been involved in genetic counseling of patients and with the application of computers in clinical genetics.

Maxine Frank Singer (1931–)

Maxine Singer was born in New York City on February 15, 1931. She received her A.B. from Swarthmore College in 1952 and Ph.D. from Yale University in 1957. Singer began her career at the National Institutes of Health (NIH) as a Public Health Service Fellow and rose through the ranks working on nucleic acid chemistry and metabolism through 1987. Her work on the biochemistry of animal viruses in the 1970s brought her into contact with the issues that she would address as cochair at the Gordon Conference on Nucleic Acids in 1973. She was instrumental in drafting a letter to the National Academy of Sciences and National Institutes of Medicine expressing concern about the safety of the new recombinant DNA technology. This public anxiety of a group of scientists combined with similar sentiments of Paul Berg eventually led to the moratorium on certain research with recombinant DNA proposed at the Asilomar Conference of 1975. She continued her involvement in both recombinant DNA research and in monitoring scientific social responsibility on the safety of recombinant DNA systems. In the 1980s, when the hysteria over the new technology had died down without ecological disaster, Singer began emphasizing the benefits to be derived from recombinant DNA applications. In the 1990s, she pointed out that the Flavr Savr tomato genetically engineered to delay softening upon ripening was little different from tomatoes derived by years of crossbreeding and was just as safe to eat.

Singer's accomplishments were recognized by numerous awards including her election to the U.S. National Academy of Sciences and National Institutes of Medicine. She has been an emeritus scientist of the NIH National Cancer Institute and the president of the Carnegie Institute in Washington, D.C., from 1988 until retiring in 2002. She supervised the First Light Program designed to interest grade school girls and boys in science.

Robert Swanson (1947–1999)

Robert Swanson made his impact on genetic engineering by providing the primary ingredient that was to allow this technology to become a major force today: investment capital. With an undergraduate degree in chemistry from Massachusetts Institute of Technology and an M.S. from MIT's Sloan School of Business and fortified with four years of experience as an investment banker with Citibank, he was aptly suited for bridging the promise of the science and the principles of business competition and return on investment on which many a biotechnology startup company founders. In 1975, Swanson found Herbert Boyer at the University of California, San Francisco, where his and Stanley Cohen's labs combined the use of the new restriction enzymes to cut DNA, leaving sticky ends with bacterial plasmids to recombine genes, inserting them into bacterial cells for the production of specific proteins. Assured by Boyer that commercial application of genetic engineering was feasible, in 1976 the two convinced Thomas Perkins of Kleiner-Perkins, a Silicon Valley venture capital firm, to advance seed money to establish Genentech. By 1978, Genentech had produced genetically engineered human insulin in bacteria and licensed the technology to Eli Lilly, a major pharmaceutical company and producer of animal (porcine) insulin for treatment of diabetes. In 1985, Genentech became the first startup biotechnology company to introduce its own biopharmaceutical product, human growth hormone. It remains the largest and most successful biotechnology company both on the business side and in scientific enterprise. Swanson was director and chief executive officer from 1976 until 1990 when he was named chairman of the board. He died in 1999.

James Alexander Thomson (1959–)

James Alexander Thomson was born in Oak Park, Illinois, and obtained his B.S. in biophysics from the University of Illinois in 1981. He earned a doctor of veterinary medicine degree in 1985 and a doctorate in molecular biology in 1988 from the University of Pennsylvania where he studied genetic imprinting in early mammalian development. This research area proved seminal in focusing his interests. After two years as a postdoctoral researcher at the Primate In Vitro Fertilization and Experimental Embryology Laboratory at the Oregon Regional Primate Research Center, he joined the Wisconsin Regional Primate Research Center of the University of Wisconsin where he developed techniques for isolation and culture of nonhuman primates, publishing the first isolation of nonhuman primate embryonic stem cells in 1995. In 1998, a group headed by Dr. Thomson reported the first isolation of human embryonic stem cells in 1998, a feat accomplished nearly simultaneously by the group of John Gearheart at the Johns Hopkins School of Medicine.

Dr. Thomson is a champion of the ethical use of embryonic stem cells for basic research into developmental biology and for therapeutic uses. He advocates for the removal of federal government restrictions on government funding of stem cell research and supports increased funding for embryonic cell research. He is the founder of Cellular Dynamics International, a biotech startup focused on differentiating embryonic stem cells into heart cells for testing heart drugs. They hope to change medicine rapidly rather than wait for the development of transplantation techniques and the resolution of myriad ethical issues. He is also scientific director for WiCell Research Institute, a research center in Wisconsin dedicated to embryonic stem cell biology studies.

J. Craig Venter (1946–)

Craig Venter was born on October 14, 1946, in Salt Lake City, Utah. A year spent in Vietnam as a medical corpsman during the war there in 1967 helped galvanize his career plans along medical lines. In 1972, he obtained his B.A. from the University of California, San Diego, where three years later, in 1975, he earned his Ph.D. in physiology and pharmacology. After a research fellowship from 1975 to 1976 in cardiovascular pharmacology at the University of California, San Diego, he moved to the State University of New York at Buffalo, where he was a faculty member in pharmacology and biochemistry until 1984 when he joined the National Institute of Neurological Disorders and Stroke at the National Institutes of Health (NIH) in Bethesda, Maryland. From his initial studies on pharmacologically important hormone receptors, Venter entered into the then-infant field of genomics, promoting a sequencing strategy called the expressed sequence tag/complementary DNA (EST-cDNA) technique to identify human genes, skipping the inactive (nontranscribed) parts of the genomic DNA sequence. The NIH raised a storm of scientific controversy when it sought to patent the EST gene fragment sequences without knowing the function of the proteins coded by the sequences. Many critics believed that patenting gene sequences would discourage research on the gene products. This criticism added to the controversy over whether genes should be patentable at all. The EST/cDNA technique coupled with the growing ability to predict the function of new sequences based on characteristic protein sequence structural and functional motifs has been useful in chromosomal mapping projects because they are large enough (300–400 base pairs) to represent unique sequences, which can be rank-ordered along the length of the chromosomes.

In 1992, failing to obtain expanded research funding for his EST sequencing project, Venter once again precipitated controversy by jumping from NIH to become the president and director of a private, nonprofit research center, the Institute for Genomic Research (TIGR) in Gaithersburg, Maryland. TIGR was allied with a biotechnology affiliate Human Genome Sciences (HGS), which was to commercialize selected sequences discovered by TIGR. Access to a cDNA database jointly maintained by TIGR and HGS and preferentially available to pharmaceutical companies supporting their research has been a bone of contention. Much of the disputed data, some 45,000 genes, has recently been made publicly available over the World Wide Web. The human genome sequence was jointly reported by Celera and the government Human Genome Research Institute in June 2001. Such debate over public and private interests has been a common feature of biotechnology. It has been at once an ethical challenge for modern society and the driving force for the application of the new knowledge for the benefit of humankind. Venter left Celera in 2002 and has been involved with several nonprofit institutes, finally merged in 2004 as the Craig C. Venter Institute. His current interest is microorganisms found in seawater. In June 2005, he founded Synthetic Genomics dedicated to the use of modified microorganisms to create ethanol and hydrogen as alternative fuels.

James Dewey Watson (1928–)

James Watson was born on April 6, 1928, in Chicago, Illinois. He entered the University of Chicago at age 15, receiving his B.S. in 1947. Harvard and the California Institute of Technology rejected his applications for graduate study (ironically, he would later serve on the faculty of both institutions), so he went to Indiana University where he earned his Ph.D. in genetics in 1950. After postdoctoral study in Copenhagen (1950–1951), Watson joined the Cavendish Laboratory in Cambridge, England, where he met Francis Crick and became involved in the search for the structure of DNA. As described in the autobiographical *The Double Helix* (Atheneum, 1968), he and Crick fashioned a model of two DNA strands wound around each other in a helical configuration held together by the pairing of DNA bases. Data for the model came from X-ray crystallography and electron microscopic studies of Rosalind Franklin and Maurice Wilkins as well as from their own work. The DNA double helix was published in 1953 and served as a cornerstone for nucleic acid biology in the infant fields of molecular genetics and biochemistry. For the far-reaching consequences of this paradigm, Watson, Crick, and Wilkins were jointly awarded the Nobel Prize in Physiology and Medicine in 1962.

Many scientific awards followed his seminal work including election to the U.S. National Academy of Sciences. Watson was a faculty member at CalTech (1953–1955) and finally at Harvard (1956–1976). He continued to make contributions to the understanding of the triplet code of DNA by which the sequences of proteins are specified. In 1968, he served as director of the Cold Spring Harbor Laboratory where he continued through the present, becoming president in 1994 and the chancellor in 2004, administering and advocating for basic research in science. When the Human Genome Project needed a prominent, forceful champion for their huge DNA sequencing project, they turned to Watson, who provided the required spark and leadership from 1988 to 1990 to get the project rolling, pointing out its potential medical applications. He once commented, "We used to think that our future was in the stars. Now we know that it is in our genes." Always outspoken and, many felt, often abrasive, James Watson has proved to be an able administrator and a positive force for the application of gene science. A staunch advocate of unfettered intellectual pursuit and the scientific method, he has been criticized for not paying enough attention to the impact of the Genome Project on political, social, or ethical issues.

6

Facts, Data, and Opinion

Statistical Data

Understanding and Acceptance of Genetic Engineering

The recombinant DNA revolution has experienced the same resistance to the introduction of new technology commonly seen in Western society throughout history. Analysts cite a variety of theories for this behavior, with a common motivation being fear of the unknown. A survey covering the period from 1983 to 2001 of the attitude of the American public toward organized science indicated that as people have become more accustomed to technological advances, they become less concerned with the rate of change and tend to feel that those changes contribute positively to their lives (Figure 6.1).

Influential factors in the recombinant DNA debate include the potential impact on world ecosystems, health, and personal control of information. Many other major controversies with implications for public health such as locating the Seabrook nuclear power plant, drilling for oil offshore or in the Arctic Wildlife Reserve, landing of the Supersonic Transport aircraft, regulating the sale of the artificial sweetener saccharin, establishing standards for genetic carcinogens, and reducing ozone destruction by fluorohydrocarbons such as Freon® were decided in the public forums of the courts, by legislation, and in public hearings of government agencies. Scientific input was only part of the process.

The genetic engineering controversy, particularly during the initial stages of the debate, has been distinct on several levels.

141

FIGURE 6.1
Attitude toward Organized Science

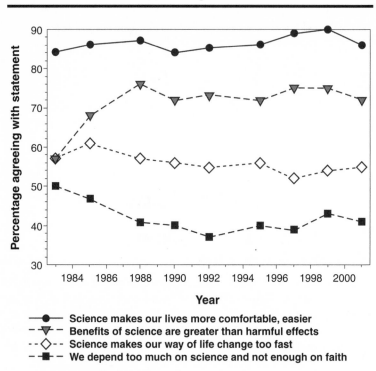

Source: Revised from National Science Board, *Science and Engineering Indicators — 2002* (Washington, DC: U.S. Government Printing Office, 2002), Appendix, Table 7-13.

Recombinant DNA technology is largely a scientific research tool, and scientists, a decided minority of the population, initially would have the most to gain from its application. As the technology matured, applications attracted business and economic interests that acted on a larger scale, consequently arousing public suspicion of the technology. The potential hazards, although widely discussed, had not been confirmed to occur, nor was the magnitude of the expected effects known. Finally, in addition to the environmental and health issues, the public debate included *ethics.* Even if personal rights of privacy of genetic information and fair treatment are guaranteed, is it *proper* to manipulate the genetic substance in the first place? Achieving consensus on such issues of faith is a daunting prospect. A survey in the United

Kingdom (Lee, Cody, and Plastow 1985) found that 70% of the respondents found genetic engineering to be "morally wrong," 62% found it "unnatural," and 27% found it "frightening." Similar responses were obtained in the United States (Hoban and Kendall 1992). The beliefs about the ethical scruples also depend on the proposed use of the genetic engineering. People are far more likely to consider the application of genetic technology permissible for medical conditions than for food enhancements or increased industrial production.

Outside of the "gut" feelings of moral rightness and trepidation toward change, a major contributor to the general public unease with genetic engineering is the highly technical nature of the arguments used to justify the safety of the technology and to assess the magnitude of the harm that could be caused. This reflects the general scientific illiteracy of the American public, particularly in the lack of understanding of the process of scientific inquiry—how scientists arrive at conclusions and what those conclusions mean. A 2005 survey by the Office of Technology Assessment (National Science Board 2002, pp. 7-10 to 7-12) concluded that 70% of adults, up from 64% in 1995, do not understand the process by which measurements are made and comparisons drawn in experimental studies to determine which of two alternative treatments is better than the other and whether the difference is significant. As for many issues, though, lack of understanding is no impediment to having strongly held opinions. Formal education, in particular, science and math education, improves understanding of the process, but still nearly one-third of respondents do not understand basic scientific proof (Table 6.1).

People's feelings about whether the benefits to be derived from genetic engineering were greater than the risks showed little correlation with educational level from non-high school graduate to baccalaureate. If anything, non-high school graduates were slightly more optimistic about the benefits than those with higher education (Figure 6.2).

Early fears about genetic engineering focused on the probability of the escape of modified organisms from laboratories causing human disease and disruption of the environment. Initially, little experimental evidence was available to drive bona fide risk analysis; even the scientists had to make quite unsophisticated estimates. As data became available and biocontainment—the development of standard workhorse organisms with drastically

TABLE 6.1
Public Understanding of the Nature of Scientific Inquiry

Level of Understanding	A	B	C	D	Sample Size
		(% of	sample)		
All adults	2	21	13	64	2,006
Formal education					
Less than high school	0	4	7	89	418
High school graduate	1	18	15	66	1,196
Baccalaureate	6	44	13	37	260
Graduate/professional	10	49	12	29	132
Science/mathematics education					
Low	0	9	12	79	1,125
Middle	3	30	16	51	530
High	7	45	12	36	352
Sex					
Female	2	20	13	65	1,053
Male	2	22	12	64	953
Attentiveness to science or technology					
High	5	34	14	47	195
Medium	3	22	13	62	946
Low	1	16	13	70	865

Respondents were presented with the following situation: "Two scientists want to know whether a certain drug is effective against high blood pressure. The first scientist wants to give the drug to 1,000 people with high blood pressure and see how many experience lower blood pressure levels. The second scientist wants to give the drug to 500 people with high blood pressure and not give the drug to another 500 people with high blood pressure and see how many in both groups experience lower blood pressure. Which is the better way to test this drug? Why is it better to test the drug this way?"

A = understands science as the development and testing of theory.
B = not A level understanding, but understands concept of experimental study, including meaning and use of a control group.
C = not B level understanding, but understands science to be based on careful and rigorous comparison.
D = does not understand science on any of the above levels.

Source: Revised from National Science Board, *Science and Engineering Indicators–1996* (Washington, DC: U.S. Government Printing Office), Appendix, Table 7-9, p. 304.

reduced ability to survive outside of the laboratory—improved, the fantastic scenarios of massive plagues faded from discussion. There remain serious questions about the safety of the large-scale deliberate release of organisms into the environment for agricultural or bioremediation applications, which are presently addressed on a case-by-case basis through the Environmental Protection Agency (EPA) and allied agencies. Surveys by the Office of Technology Assessment published in 1987 indicate that public opinion is less than mollified. Respondents made little distinction

FIGURE 6.2
Public Assessments of Genetic Engineering by Educational Status

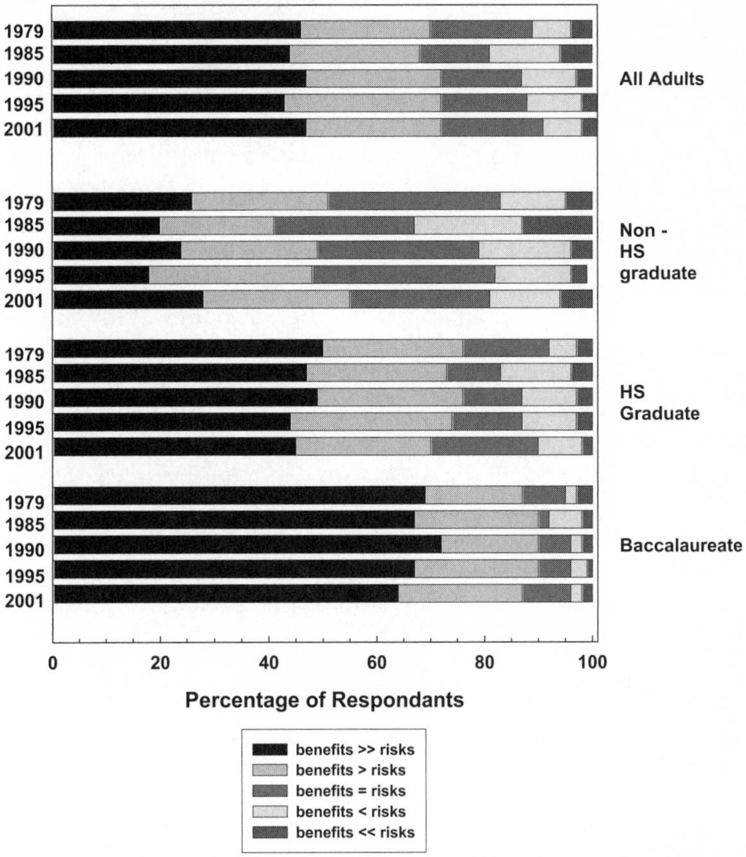

Source: Revised from National Science Board, *Science and Engineering Indicators — 2002* (Washington, DC: U.S. Government Printing Office, 2002), Appendix, Table 7-23, pp. 1072–1073.

between specific dangers that might be expected from the use of genetically altered organisms in the environment (Office of Technology Assessment 1987). When presented with approving agricultural use of an organism with no risk to humans but with increasingly remote possibilities of losing some local species of plant or fish ranging from unknown up to one in a million, a substantial percentage of respondents (18%) would not approve even at the lowest risk ($1:10^6$). Significantly, having available some quantitative measure of risk, such as 1:1000 or 1:10,000 provided a basis for a decision. Forty-six percent would not approve the use of the organism if the risk were unknown even if it were "very remote." The specific use of the organisms affected the likelihood of public approval because release of disease- and frost-resistant crops or "oil-eater" bacteria was more likely to be approved than more effective biopesticides or larger game fish.

While people wanted to be consulted on whether or not genetically engineered organisms should be released into the environment, they were more comfortable delegating the actual decisions on whether commercial firms should be allowed to release genetically modified organisms on a large scale to a government agency or secondly to an external scientific body. The distribution of opinion was similar across different political affiliations. Interestingly, only 5% or less of respondents believed that the public at large or communities should decide.

Surveys conducted over 10 years later asking most of the same questions have been conducted in other countries including New Zealand and Japan (Macer 1998), the European Union (BEPCAG 1997), and Canada (Eisiedal 1997). Despite the differing extent of involvement in biotechnology, different cultural traditions, and the intervening 10+ years since the U.S. survey, the opinions of the sampled populations are remarkably consistent in their basic mistrust of biotechnology, fuzzy public understanding of questions of relative risk-to-benefit, apprehension about effects on the environment, and wanting to be consulted in deciding what kind of work should be allowed. There remains a general lack of confidence in the effectiveness of government regulation and in the motivations and accountability of industry. People are far more willing to see genetic engineering used in health-related ways than for food production, and all viewed the potential for human cloning research negatively.

Genetic Engineering as a Business

Genetic engineering has been a vastly enabling technology at both the level of being able to do things that were previously impossible or completely novel and in expediting and improving the capabilities of current practices. Many believe that this success and rapid development, especially in the United States, was primarily because obstacles to its commercialization were removed at an early stage. The General Electric Company/Chakrabarty patent decision from the Court of Customs and Patent Appeals was a landmark determination. It allowed the patenting of bacteria engineered to metabolize oil spills during the early 1970s as a new form of life. Previously, certain types of plants were the only form of life that could be patented. Eventually upheld by the Supreme Court in 1980, the Chakrabarty patent plus an additional patent to Stanford University (Drs. Stanley Cohen and Herbert Boyer) on gene splicing methodology issued the same year laid the legal groundwork for the biotechnology industry. In 1987, the first patent was granted to Harvard University on a genetically engineered mammal, a mouse designed to be highly susceptible to tumor formation to aid in the development of anticancer therapies. A patent grants an exclusive right for a specified period of time to challenge the use of that technology or product without permission of the holder of the patent. Each country has its own set of rules governing the protection of intellectual property by patents or copyrights, although international agreements have established some common standards.

Patents are currently allowed in the United States for genetically modified microbes, cells, animals, and plants, whether produced by recombinant DNA technology or by traditional breeding practices, although some restrictions are under discussion. This view is challenged by a number of activists and some countries such as India. New techniques for manipulating genes and inserting them into organisms can also be protected. A gene itself may be patented (the use of the gene and its sequence), but only if the function of the gene is known, which prevents patenting of unknown DNA sequences in hopes of figuring out what they are later. The rules governing patents in biotechnology evoked much controversy when they were first proposed, and they remain a complex subfield of patent law. Although not a guarantee of exclusivity, patent protection attracted financial investment to the new biotechnology companies, which gave them the money and therefore the time to apply the new genetic technology to product development.

Even in the age of colossal multinational corporations, a technique, an idea, a few test tubes, some agar plates, and a patent pending can put a person into the genetic engineering business. Usually such operations begin in the university research laboratory of a faculty member, but more than a few started in a garage or vacant warehouse. The uses of world biotechnology grew exponentially from 1969 to 1984; since 1983, the growth has simply been explosive. The United States has tended to dominate the distribution of biotech companies in the world, partly because of the availability of investment capital and technical expertise and partly because of favorable intellectual property laws. In 1988, 469 of the worldwide 1,036 biotechnology companies were in the United States. The common technology has spurred a worldwide network of strategic alliances (Table 6.2).

By 2003, there were 1,473 biotechnology companies in the United States alone. They are distributed primarily among states that have made a conscious effort to attract the new technology (Table 6.3).

Most are small, with 58% of the companies employing fewer than 50 people and less than 12% with more than 500 workers. They are distributed thus: 1–10 employees, 25.7%; 11–50 employees, 32.9%; 51–500 employees, 30.9%; 501–2,500 employees, 5.7%; 2,501–15,000 employees, 3%; and >15,000 employees, 1.9% (*A Survey* . . . 2003). They also tend to be concentrated in metropolitan areas. The top 10 states in total biotechnology research and development funding in the *Science and Engineering Indicators*—*2002* survey were California, Michigan, New York, Texas, Massachusetts, Pennsylvania, New Jersey, Illinois, Wisconsin, and Maryland (National Science Board 2002, Text Table 4-11, p. 196).

The number of biotechnology firms in the United States has remained fairly constant since 1994 (1,473 in 2003). Sales increased almost fourfold as their development streams matured into saleable products, recording an aggregate of more than $28.4 billion in sales in 2003. The number of employees over this same

TABLE 6.2
Worldwide Strategic Alliances of Biotechnology Companies (1990–2000)

United States–Europe	525
United States–Japan	82
United States–other	71
Europe–Japan	37
Europe–other	49
Japan–other	6
Intra-United States	629
Intra-Europe	147
Intra-Japan	7
TOTAL	1,553

Source: National Science Board, *Science and Engineering Indicators*—*2002* (Washington, DC: U.S. Government Printing Office), Appendix, Table 4-12, p. 208.

TABLE 6.3
Location of U.S. Biotechnology Companies, 2004

California	420
Massachusetts	193
North Carolina	88
Maryland	84
New Jersey	77
New York	66
Texas	64
Georgia	63
Pennsylvania	63
Washington	42
Florida	33
Connecticut	29

Source: Modified from Ernst and Young, LLP, *America's Biotechnology Report: Resurgence 2004.* Available: *BIO* Web site: http://www.bio.org/speeches/pubs/er/statistics.asp (04-21-06).

period nearly doubled, again reflecting the maturing of the industry, with a total of 198,300 employees by the end of 2003. These employment figures do not include those working in government laboratories, academic scientists, or state biotechnology centers. Therapeutics and healthcare account for the focus of more than 75% of the companies (*A Survey . . . 2003*).

This late-20th-century "cottage industry" spawned a new generation of biotechnology investment capitalists to provide the hundreds of millions of dollars of financial support needed to turn the golden idea or technology into a product. In 1992, compared with total sales of $5.9 billion, biotechnology firms spent $4.9 billion on research, roughly equivalent to IBM's research and development budget that year. More than 90% of these new ventures failed, like many new businesses, most within the first 5 years, because they didn't convert their ideas into moneymaking products fast enough. Only 18% of 225 public biotechnology firms were profitable in 1995. Good scientists frequently are not good business managers. The 1990s brought a slowdown in biotechnology investing because the fantastic forecasts failed to materialize and the stock market general technology bubble burst. Nevertheless, by persevering or by shrewd alliances with established industries, many successful biotechnology-based companies have emerged. Most are connected with healthcare, although agricultural applications continue to expand. Major corporations have also invested or acquired interests in biotechnology as they recognized the need for such processes or investments (Table 6.4).

TABLE 6.4
Some Major Corporations Investing in Biotechnology

Abbott Laboratories	General Foods Corp.
Allied Chemical Corp.	Genzyme Corp.
Allied Signal, Inc.	GlaxoSmithKline
American Cyanamide Co.	Hercules Research & Development
American Home Products	Hoffman-LaRoche, Inc.
American Hospital Supply Corp.	Johnson & Johnson
Amgen	Kimberly-Clarke
Amoco Corp.	Life Technologies, Inc.
Ares-Serono Laboratories	Litton Bionetics, Inc.
Aventis-Behring	Lubrizol Enterprises
BASF	Merck and Company, Inc.
Baxter Healthcare Corp.	Miles Laboratories, Inc.
Becton Dickinson and Co.	Miller Brewing Company
Berlex Corp.	3M
Biogen-Idec	Monsanto Agriculture Company
Bio-Rad Laboratories	National Distillers & Chemical Corp.
Boehringer-Ingleheim Corp.	New England Nuclear Corp.
Boehringer-Mannheim Corp.	Novartis Pharmaceutical Corp.
Bristol-Meyers Squibb Corp.	Olin Corp.
Celanese Research Co.	Organon
Chiron Corp.	Ortho Pharmaceutical Corp.
Corning Glassworks	Pennwalt Corp.
Del Monte USA	Pfizer, Inc.
Diamond Shamrock Biotechnology Research	Phillips Petroleum Co.
Dow Chemical Co.	Proctor and Gamble Co.
Dupont Pioneer-Hi-Bred	RJR Nabisco, Inc.
E. I. du Pont de Nemours Co.	Rohm & Haas Co.
Eastman Kodak Co.	Rorer Group Inc.
Ecogen Inc.	Schering-Plough Corp.
Elan Corp.	Standard Oil Co.
Eli Lilly & Co.	Texaco Research Center
Exxon	Universal Foods Corp.
FMC Corp.	Weyerhauser Co.
Genentech	W. R. Grace & Co.
General Electric Co.	Wyeth Laboratories

Source: S. S. Brown, *Opportunities in Biotechnology Careers* (Lincolnwood, IL: NTC Publishing Group, 1994), pp. 138–141.

Their individual research and development expenditures extended into the hundred of millions of dollars in 1995 (Table 6.5). Note that some companies have grown in worth (Market Capitalization) with relatively low research and development expenditures. This type of success depends both on the type of products and the company's financial situation.

TABLE 6.5
Top 48 Biotechnology/Pharmaceutical Companies[a]
Worldwide by 2003 Research and Development Expenditures

Company	R&D (millions US$)	Market Capitalization (billions US$)[b]
Pfizer, USA	$3,983.6	$194.1
GlaxoSmithKline, UK	$2,791.0	$143.4
Johnson & Johnson, USA	$2,616.6	$192.6
Novartis, Switzerland	$2,098.2	$115.5
AstraZeneca, UK	$1,927.8	$79.2
Merck, USA	$1,775.4	$64.4
Eli Lilly, USA	$1,312.9	$64.3
Bristol-Myers Squibb, USA	$1,273.1	$49.1
Wyeth, USA	$1,169.5	$61.5
Sanofi-Synthelabo, France	$927.3	$231.6
Amgen, USA	$924.8	$138.9
Schering-Plough, USA	$820.6	$32.9
Allergan, USA	$425.7	$18.1
Novo Nordisk, Denmark	$396.8	$31.1
Eisai, Japan	$311.2	$17.2
Altana, Germany	$290.2	$14.5
Millennium Pharmaceuticals, USA	$272.9	$4.7
Serono, Switzerland	$272.3	$16.5
Chiron, USA	$218.6	$12.7
Genzyme, USA	$187.3	$28.5
Elan, Ireland	$142.1	$5.7
Forest Laboratories, USA	$137.7	$24.3
Biogen Idec, USA	$130.4	$24.3
Shire Pharmaceuticals, UK	$128.4	$9.5
Sepracor, USA	$123.0	$11.6
Teva Pharma Industries, Israel	$119.3	$36.7
Human Genome Sciences, USA	$102.4	$2.7
Cephalon, USA	$95.1	$4.5
Gilead Sciences, USA	$92.1	$33.3
MedImmune, USA	$87.3	$11.6
Vertex Pharmaceuticals, USA	$77.9	$2.0
Exelixis, USA	$71.3	$1.0
Intermune, USA	$67.0	$0.7
Incyte, USA	$64.9	$1.1
Ivax, USA	$60.5	$9.1
Celgene, USA	$60.1	$12.2
Mylan Laboratories, USA	$56.3	$8.1
Watson Pharmaceuticals, USA	$55.5	$5.9
Medarex, USA	$53.3	$1.5
OSI Pharmaceuticals, USA	$52.0	$3.6
Barr Pharmaceuticals, USA	$51.0	$9.3

continues

TABLE 6.5 continued

Company	R&D (millions US$)	Market Capitalization (billions US$)[b]
ImClone Systems, USA	$49.6	$5.0
Regeneron Pharmaceuticals, USA	$49.5	$0.6
ICOS, USA	$47.9	$2.5
Abgenix, USA	$46.2	$1.2
Neurocrine Biosciences, USA	$45.0	$2.5
Affymetrix, USA	$36.8	$6.1
Amylin Pharmaceuticals, USA	$35.6	$3.0

[a] Companies with publicly traded stock
[b] September 2005 intraday market capitalization

Sources: DTI 2004 R&D Scoreboard 700 Top UK and 700 International Companies by R&D Investment, part 2. Company Data http://www.innovation.gov.uk
Market Capitalization – Yahoo Finance

Total spending of pharmaceutical and biotechnology companies (not all necessarily related to genetic engineering) on research and development climbed up to 1995 and then leveled off (Figure 6.3), reflecting the investment downturn of the late 1990s.

Over the period from 1981 to 1992, a national survey showed that research and development expenditures in several categories (motor vehicles; office and computing; and food, beverages, and tobacco) leveled off, whereas medicines and drugs (including biotechnology) nearly tripled (Figure 6.4). According to the Pharmaceutical Research Manufacturers Association (PhRMA) Foundation (Washington, D.C.), U.S. companies invested $33.2 billion in research and development (17.7% of domestic sales), more than the National Institutes of Health (NIH) and the international pharmaceutical industry combined. In 2004, these figures rose to $38.8 billion (18.8% of domestic sales).

It is becoming increasingly difficult to separate biotechnology and the pharmaceutical sector because recombinant DNA technology is becoming so pervasive in that industry. U.S.-company-financed research and development in pharmaceuticals continues to exceed that of other countries by a wide margin (Table 6.6). There are 155 Food and Drug Administration (FDA)–approved biotechnology derived drugs and vaccines on the mar-

FIGURE 6.3
Company Funding of Pharmaceutical Research and Development

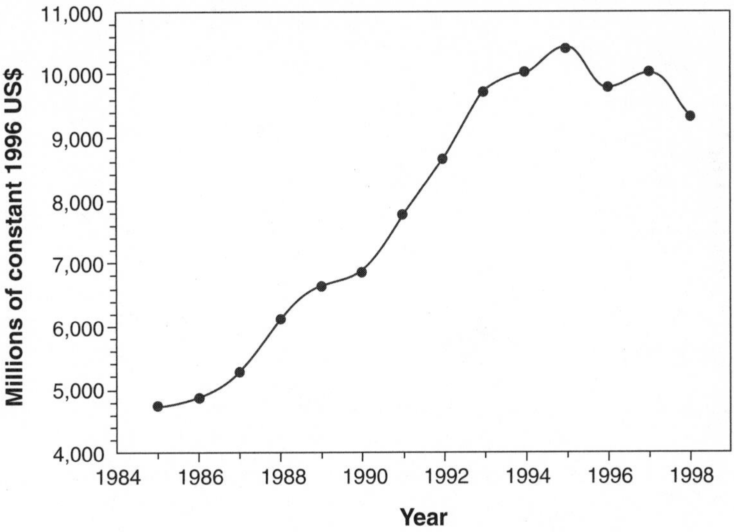

Source: Revised from National Science Board, *Science and Engineering Indicators—2002* (Washington, DC: U.S. Government Printing Office, 2002), Appendix, Table 4-29, p. 810.

ket and more than 370 currently in clinical trials targeting more than 200 diseases.

These figures may actually underestimate the investment of the industrial sector because it is difficult to separate private industry:government coalitions in such countries as Japan and Cuba in which the economic model differs from the United States. The U.S. federal government investment in biotechnology research is considerable, $4.299 billion in Fiscal Year (FY) 1994, of which the lion's share, 41%, was devoted to improving medicine and health-related programs. Much of the funding goes to universities to provide both the basic groundwork to advance the science and to stimulate application of new techniques through research foundations (39%) and to support technology infrastructure in general (8%). Agriculture, manufacturing, environment, and energy resources receive much less governmental support (12% total). Some people believe that these fields have potentially as great if not greater social impact as healthcare in a world in which food and

FIGURE 6.4
Manufacturing Industry Research and Development Performance, 1981–1997

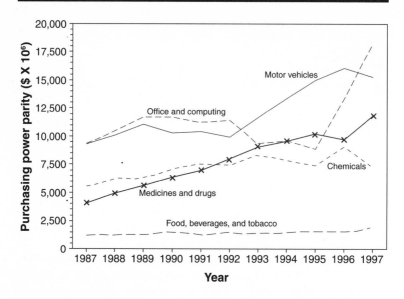

Source: Revised from National Science Board, *Science and Engineering Indicators — 2002* (Washington, DC: U.S. Government Printing Office, 2002), Appendix, Table 6-9, p. 1033.

TABLE 6.6
Company-Financed Pharmaceutical Research and Development Ranked by Country for 1995

United States	36%
Japan	19%
Germany	10%
France	9%
United Kingdom	7%
Switzerland	5%
Sweden	3%
Italy	3%
Other countries	8%

Source: Revised from Centre for Medicines Research International, 1997 figures. In *Parexel's Pharmaceutical R&D Statistical Sourcebook* (Waltham, MA: Parexel International Corp., 1997), p. 167.

energy supplies dwindle as population and demand increases. Research in several of these areas, particularly agriculture and manufacturing, is dominated by private industry in the United States (Figure 6.5).

The Cycle of Ideas

It is useful to consider today's trends in the context of historical developments. According to Nola Masterson, Ph.D., of Science Futures in San Francisco (Wong 2005), revolutionary ideas or processes that profoundly impact society such as computer

FIGURE 6.5
Federal Investment in Biotechnology Research—Fiscal Year 1994

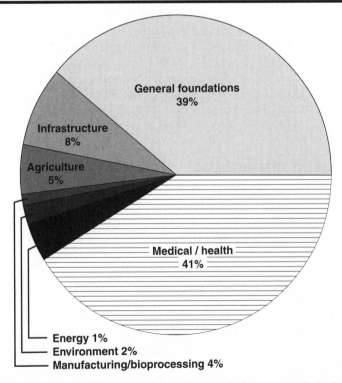

Source: From National Science and Technology Council, Biotechnology Research Subcommittee, *Biotechnology for the 21st Century: A Report from the Biotechnology Research Subcommittee, Committee on Fundamental Science* (Washington, DC: Biotechnology Research Subcommittee, 1995), p. 2.

technology, automobiles, and biotechnology tend to follow a generational cycle that operates over 50 years:

- The first decade lays down the foundation for the technologies.
- The second decade yields a small number of products based on this methodology.
- The third decade produces a flood of innovative products based on refining and broadening its core methodologies.
- The fourth decade finds products incorporated into daily life, and the processes that generated them are fully integrated into industry.
- The fifth decade ushers the formerly new technologies into mainstream industry, and the revolution that gave birth to them becomes part of history.

By this timeline, the chemical industry is in its fourth decade, as plastics and polymers are commonplace. Biotechnology is beginning its third decade in which a flood of products is making its way into clinical trials and onto the market. It is also shifting its research focus into cellular and systems biology to harness the massive amount of knowledge gained in genomics and proteomics for cell-based therapies and stem cell research.

Medical Applications

The top 25 publicly traded biotechnology companies worldwide (Table 6.5) are heavily invested in medical and health product development. This position is driven by the ready application of genetic engineering to the production of scarce biomolecules, the relatively high rate of return on investment for pharmaceuticals, and a low profile on environmental impact because the recombinant organisms are controlled entirely within the production facilities. Those companies diversifying into other areas such as agriculture and bioremediation have encountered the ecological issues surrounding release of engineered organisms into the environment and a significant public backlash against the technology. The FDA approved the first recombinant biopharmaceutical agent, human insulin (Humulin) from Eli Lilly, in 1982. These recombinant products are mostly protein hormones or growth factors (of human sequence to avoid immunological reaction), replacement enzymes (blood clotting factor VIII), or antibodies designed to neutralize or target cell types for cancer therapy. The top-selling biotech drug worldwide is Amgen's erythropoetin, an activator of blood cell production, ranking 15th in the worldwide drug sales overall for 1996. Four recombinantly engineered drugs were among the top 50 drugs sold in 1997. To put it in perspective, though, biopharmaceuticals account for only about 10% of 1996 U.S. drug sales (Dibner 1998).

Gene therapies for cancer, cystic fibrosis, and enzyme replacement are also being developed. In contrast to the initial trials in 1990 of adenosine deaminase (ADA) replacement in ADA-deficient immunocompromised individuals which were designed as a proof of concept for gene therapy by W. F. Anderson, the diseases being targeted now afflict significant numbers of people. For both technical and ethical reasons, *no* permanent genetic repairs in which the defective gene is replaced or otherwise inactivated in the reproductive cells, transmitting the modification to future generations, are being advanced. Only the individual being treated (presumably) will be affected by the therapy. Targeting and the robustness and duration of the therapeutic effect will be carefully monitored. The difficulties of this approach are legendary and the ability of such treatments to compete with more conventional pharmacologic avenues, when they can be developed, will decide the success of these companies.

A concern of both proponents and opponents of gene therapy is the risk of abuse of the technology. Although there is general agreement that gene therapy should be subject to societal controls, defining what constitutes abuse will be difficult. The NIH Recombinant Advisory Committee (RAC), charged with approving gene therapy protocols, is holding a series of conferences and open forums designed to bring the public into the debate on what sorts of treatments are medically justified. The NIH RAC member and bioethicist Eric Juengst from Case Western Reserve University in Cleveland, Ohio, observed, "Just about any enhancement could be packaged as therapy in some context." His thoughts are echoed by Sheila Rothman of Columbia University in New York City, who says, "Powerful forces are available to promote enhancement, and medical treatments can quickly become enhancements." A hypothetical case in some discussions—would a gene therapy solution to male pattern baldness be a "therapy" or an "enhancement"?

Agriculture and Manufacturing

Agricultural products register the largest commercial impact outside of medically related uses of genetic engineering. In 1992, out of a total of $5.9 billion in receipts, agribiotech accounted for sales of $184.5 million. In 2002, genetically engineered crop sales in the United States were $20 billion (Runge and Ryan 2003). Large chemical and pharmaceutical companies such as Monsanto, Dupont, and Searle have diversified by investing in agricultural applications of genetic engineering through interests in other companies like Collagen, Biogen, Genentech, Genex, and Biotechnica International (see Table 6.4). These companies deal in proprietary technologies rather than the commodities themselves: they provide the engineered seed and supporting treatments such as herbicides. Condensation of the industry through mergers has narrowed the field since 2000. Syngenta, Bayer, Monsanto, DuPont/Pioneer-HiBred, Dow, and BASF had worldwide sales in 2002 of seed and supporting treatments of $28 billion (Runge and Ryan 2003).

Although scientific traditional breeding of plants through controlled pollen transfer has flourished for hundreds of years, the in vitro propagation of plants from clones derived from single cells to develop desired qualities was a 20th-century phenomenon. Single plant cell cloning in its original sense does not invoke gene manipulation. Genetic engineering, in the form practiced at the molecular level for bacteria, fungi, and animal cells, was slowed by the initial lack of suitable vector systems for transfecting plant cells and difficulty in penetrating the tough plant cell wall. Some types of crops such as rice, bananas, and cereals that would impact food production for the worker population in developing nations have been recalcitrant to molecular manipulation, but they are now yielding to improved technology. The full-scale use of engineered plants, unlike most recombinant bacterial or fungal production applications, necessarily involves release of large numbers of recombinant organisms over millions of acres of cropland. Although they cannot walk, swim, or fly away, plants can disperse their genetic material (pollen and seeds) over considerable distances, potentially interbreeding with related weed plants. Transgenic trees could spread engineered sawmill-friendly characteristics such as low lignin content from cultivated tree farms to offspring of wild populations, which could affect their ability to compete and survive in unprotected areas in mixed culture with other species. Pest resistance to specific recombinant strategies can also develop, potentially causing disastrous crop failures when pest populations surge. Environmentally tolerant plants could move into new ecosystems disrupting the growth of indigenous species. As a result, the ecologic consequences of agriculturally based genetic engineering have attracted much attention. This has been a prime target for protest groups and governmental regulation. Public misgivings about pest management with genetic engineering have drastically slowed deployment of the technology. Dr. R. Jones Cook of the U.S. Agriculture Research Service at Washington State University, Pullman, Washington, reflects the climate in his statement, "[the] question of social acceptance of pest management with transgenes . . . remains a deterrent to plant biotechnology" (American Association for the Advancement of Science Annual Meeting, Philadelphia, PA, February, 1998).

The approach for regulating genetic engineering in the United States was to expand the auspices of existing regulatory agencies and to provide supplemental legislation. The result is a complex network of interconnected responsibilities. Table 6.7 outlines the regulatory responsibilities for reviewing planned introductions of genetically modified organisms, of which plants are a part.

These regulations cover basic academic research and commercial releases. Commercial biotechnology products also have to be approved by the appropriate government agency(s) depending on the intended use of the product (Table 6.8).

From 1987 through 2005, 13,357 field trials (releases) have been approved by the Animal and Plant Health Inspection Service (APHIS) of the United States Department of Agriculture (USDA) (Table 6.9). The relative proportions of trials for types of modifications have remained relatively constant over the years.

The USDA has 120 days from the time of submission of a formal proposal for field release of genetically modified organisms to review and approve the proposal. After a number of years of testing with no evidence of ecological impact, companies working with genetically engineered plants can request that APHIS determine that there is "no potential for plant pest risk" and certify the release of the tested varieties. Table 6.10 reveals that in 1996–1997, a number of engineered plants that had initially been ecologically suspect in principle showed no evidence under the particular conditions they were tested that they would become

TABLE 6.7
Responsibilities for Reviewing Planned
Introductions of Genetically Modified Organisms

No release in environment (contained)	
Federally funded	Funding agency[a]
Non-federally funded	NIH or S&E voluntary review, APHIS[b]
Foods/food additives, human drugs, medical devices, biologics, animal drugs	
Federally funded	FDA[d], NIH guidelines and review
Non-federally funded	FDA[d], NIH voluntary review
Plants, animals, animal biologics	
Federally funded	Funding agency[a], APHIS[b]
Non-federally funded	APHIS[b], S&E voluntary review
Pesticide Organisms	
Genetically Engineered	
Intergeneric	EPA[c], APHIS[b], S&E voluntary review
Pathogenic intrageneric	EPA[c], APHIS[b], S&E voluntary review
Intrageneric nonpathogen	EPA[c], S&E voluntary review
Nonengineered nonindigenous pathogens	EPA[c], APHIS[b]
Indigenous pathogens	EPA[c], APHIS[b]
Nonindigenous nonpathogens	EPA[c]
Other uses (microorganisms) released in the environment	
Genetically engineered intergeneric organisms	
Federally funded	Funding agency[a], APHIS[b], EPA[c]
Commercially funded	EPA, APHIS, S&E voluntary review
Pathogenic source organisms	
Federally funded	Funding agency[a], APHIS[b], EPA[c]
Commercially funded	APHIS[b], EPA[c] (if nonagricultural use)
Intrageneric combination of nonpathogenic source	EPA Report
Organisms	
Nonengineered organisms	EPA Report[*], APHIS[b]

[*] = Lead agency.

[a] = Review and approval of research protocols conducted by NIH, S&E, or National Science Foundation.

[b]APHIS = Animal and Plant Health Inspection Service (involved when microorganism is a plant or animal pathogen or regulated by permit).

[c]EPA = Environmental Protection Agency—jurisdiction for plots of more than 10 acres.

[d]FDA = Food and Drug Administration—reviews federally funded environmental research only when it is for commercial purposes.

NIH = National Institutes of Health.

S&E = U.S. Department of Agriculture, Science, and Education.

Source: From 51 Federal Register, Table II. Office of Science and Technology Policy Federal Registration Notice on a Coordinated Framework for Regulation of Biotechnology (1986), p. 23305.

TABLE 6.8
U.S. Federal Government Agency Responsibilities for
Approval of Commercial Biotechnology Products

Food/food additives	FDA[a], FSIS[b]
Human drugs, medical devices, and biological products	FDA
Animal drugs	FDA
Animal biological products	APHIS
Other contained uses	EPA
Plants and animals	APHIS[a], FSIS[b], FDA[c]
Pesticide organisms released into environment (all)	EPA[a], APHIS[d]
Other uses (microorganisms)	
1. Intergenera(ic) combination	EPA[a], APHIS[d]
2. Intragenera(ic) combination:	
(a) pathogenic source organism:	
agricultural use	APHIS
nonagricultural use	EPA[a,e], APHIS[d]
(b) no pathogenic source organisms	EPA Report
nonengineered pathogens	
(i) agricultural use	APHIS
(ii) nonagricultural use	EPA[e], APHIS[d]
nonengineered nonpathogens	EPA Report

[a] = Lead agency.
[b]FSIS = Food Safety and Inspection Service (under the assistant secretary of agriculture for Marketing and Inspection Service).
[c] = FDA is involved in relation to food use.
[d]APHIS = Animal and Plant Health Inspection Service (involved when microorganism is a plant or animal pathogen or regulated by permit).
[e] = Only for significant new use (proposed new rule, 1989).
FDA =F ood and Drug Administration.
EPA = Environmental Protection Agency.

Source: From 51 Federal Register Chart I. Office of Science and Technology Policy Federal Registration Notice on a Co-ordinated Framework for Regulation of Biotechnology (1986), p. 23304.

"weedy" or transfer their modified genetic material to native plants. Many more such certifications have been allowed in a continuing process of determinations (updates at http://www.aphis.usda.gov). Although this does not prove that the plants are safe in all environments, the initial apprehension turned out to be ill-founded. Critics insist that testing should be carried out in each different area because the ecologies differ.

Rather than writing new regulations specifically for genetically engineered products, the U.S. government has extended coverage with the same legal statutes that protect agriculture and

TABLE 6.9
Categories of Field Releases (APHIS) 1987–2005

Enhancement	Total Releases (1987–2005)	% total	% (2005)
Herbicide tolerance	3,624	27.2	32
Insect resistance	3,148	23.5	20
Product quality	2,334	17.5	19
Virus resistance	1,241	9.3	1
Agronomic properties	1,069	8.0	14
Fungal resistance	657	4.9	3
Marker gene	557	4.2	5
Bacterial resistance	107	0.8	—
Nematode resistance	17	0.1	—
Other	603	4.5	5
Total	13,357	100	

Source: Revised from Animal and Plant Health Inspection Service (APHIS) Web site; http://www.aphis.usda.gov/ (cited 05/02/05).

TABLE 6.10
No Potential for Plant Pest Risk Determinations (APHIS) 1996–1997

Product	Company	Date
Insect-resistant corn	Northrup King Company	January 1996
Herbicide-resistant cotton	Dupont Agricultural Products	January 1996
Male-sterile corn	Plant Genetic Systems (America), Inc.	February 1996
Altered ripening tomato	Agritope, Inc.	March 1996
Colorado potato-beetle-resistant potato	Monsanto Agricultural Company	May 1996
Virus-resistant squash	Asgrow Seeds	June 1996
Herbicide-tolerant soybean	AgroEvo	July 1996
Virus-resistant papaya	Cornell University and University of Hawaii	September 1996
Insect-resistant corn	DeKalb Genetics Corporation	March 1997
Insect-resistant cotton	Calgene, Inc.	April 1997
High oleic acid soybean	Dupont Agricultural Products	May 1997
Insect-resistant corn	Monsanto Agricultural Company	May 1997

[a]APHIS = Animal and Plant Health Inspection Service. Potential for plant pest risk is a measure of the chance that the genetically altered organism will crossbreed with closely related wild relatives to produce a weed plant with the new characteristic and significant growth or propagation advantage.

Source: Revised from Animal and Plant Health Inspection Service (APHIS), *BSS Biotechnology Update* (November 1997). Available: http://www.aphis.usda.gov/bbep/bp/newsletter.html (cited 05/02/05).

the nation's food supply from pests and contamination from conventional sources. Those who would have genetically engineered organisms and produce singled out for special regulation have criticized this policy decision. The statutes and the U.S. Legal Code comprise a very large document that has been revised and amended over the years. It is most readily accessed and searched via the Internet (e.g., http://www.law.cornell.edu/uscode/index.html and other sites). The indicated sections of the following statutes (Title I U.S. Code I Section; Title 7 = Agriculture; Title 21 = Food and Drugs) are considered applicable to USDA-regulated biotechnology (51 Federal Register, Office of Science and Technology Policy Federal Registration Notice on a Coordinated Framework for Regulation of Biotechnology [1986], p. 23339), although there is no special mention of biotechnology in the documents: Virus-Serum Act of 1913 (21 U.S.C. 151–158); Federal Plant Pest Act of May 23, 1957 (7 U.S.C. 150aa–150jj); Plant Quarantine Act of August 20, 1912 (7 U.S.C. 151–164, 166, 167); Organic Act of September 21, 1944 (7 U.S.C. 147a); Federal Noxious Weed Act of 1974 (7 U.S.C. 2801 et seq.); Federal Seed Act (7 U.S.C. 155 et seq.); Plant Variety Protection Act of 1930 and 1970 (7 U.S.C. 2321 et seq.); Federal Meat Inspection Act of 1907 (21 U.S.C. 601 et seq.); and the Poultry Products Inspection Act of 1957 (21 U.S.C. 451 et seq.). This is not an exhaustive list of applicable statutes. Modifications are made as technology and new issues develop. Keeping up with constantly changing biotechnology regulation is complex and time-consuming. The BioTechnology Permits Home Page (http://www.aphis.usda.gov/BBEP/BP/) is a useful source for information on regulations pertaining to agricultural biotechnology and links to other governmental agencies and their regulations.

Genetic engineering of foodstuffs is an even more complex and emotionally involved issue than the environment. According to University of Wisconsin sociologist Frederick Buttel, food biotechnology is responsible for 90% of the controversy, although medical biotechnology accounts for 90% of the products of genetic engineering. The human cloning and stem cell controversies have temporarily shifted the balance of discussion since that statement was made. Debate over the use of recombinant DNA technologies in the production of food includes not only environmental and consumer safety issues but the right to choose whether one uses a product produced in a certain way ("synthetic" vs. "natural"). In response to this debate, two bills were introduced into the U.S. House of Representatives (H.R. 2084 and H.R. 2085) that would require the distinct labeling of milk and milk products derived from cows treated with synthetic bovine growth hormone to stimulate milk and meat production intended for human consumption and the development of technologies to verify the source of the bovine hormone in foods.

H.R. 2084, the Bovine Growth Hormone Milk Act, introduced by Representative Bernard Sanders (I-VT) and 27 other representatives on July 20, 1997, imposes labeling requirements for milk and milk products produced from cows treated with synthetic bovine growth hormone. It directs the development of a test specific for the synthetic hormone and also requires that the USDA pay a lower subsidy price for milk and milk products produced with the aid of synthetic growth hormone. H.R. 2085, the Bovine Growth Hormone Milk Act (also introduced by Sanders and the 27 other representatives on July 20, 1997), imposes the same requirements as H.R. 2084 without requiring the USDA to pay a lower price for milk and milk products produced with the aid of synthetic growth hormone.

In response to the perceived impact of the growing influence of agribusiness and the unknowns of genetic engineering technologies, legislation has been introduced to address the issues confronting farmers and developing nations:

H.R. 4812: The Genetically Engineered Crop and Animal Farmer Protection Act of 2002 sets forth a Farmer's Bill of Rights, including the right of farmers to save seeds, to be informed of the risks of using genetically engineered crops, and not be held liable for effects of genetically engineered crops to others; it also prohibits genetic engineering designed to produce sterile seeds ("terminator" technology) and loan discrimination based on choice of seeds.

H.R. 4813: The Genetically Engineered Food Safety Act of 2002 requires that all genetically engineered foods follow the FDA's current food additive guidelines, phasing out antibiotic selection markers. It prohibits introduction of known allergens, authorizes the FDA to independently test food safety of genetically engineered products, and opens a public comment period after a safety application is completed.

H.R. 4814: The Genetically Engineered Food Right to Know Act of 2002 requires the labeling of all foods that contain or are produced with genetically engineered material enforced by periodic FDA testing and authorizes voluntary "no genetic engineering" labels.

H.R. 4815: Real Solutions to World Hunger Act of 2002 would offer incentives and protections to help developing nations by establishing an international sustainable agriculture research fund, allow U.S. firms to market to developing countries only the genetic engineering technology that has been certified safe for use in the United States, and make it mandatory for biotech companies to allow developing nations to license genetic engineering technology that is appropriate for the nations' peoples to produce food. A trust fund would be set up to fund these activities, financed by a tax on the biotechnology companies selling genetically engineered seeds.

H.R. 4816: Genetically Engineered Organism Liability Act of 2002 would protect farmers by assigning all liability arising from the consequences of the use of genetically engineered crops, such as ecological damage or contamination of neighboring non–genetically engineered fields, to the creator of the technology.

Such legislation faces an uphill battle from agribusiness, which will contest demonization of a technology that they feel has been proven to be safe and effective. Agribusiness is concerned about the punitive nature of this legislation and asks why these businesses are being held to a higher standard than other modes of food production. The issues are a complex mixture of antitechnology philosophy and a growing distrust of the industrialization of farming and food production by huge multinational corporations that have no boundaries and no loyalties except to their financial bottom line. Of the examples cited only the Growth Hormone Act has been passed into law, although components of the other legislation may still be integrated into other legislation.

Despite the confusing web of regulation, the perceived future returns on investment of genetically engineered crops have kept entrepreneurial interest high. As shown in Table 6.11, in addition to the expected agricultural and food-based players, large chemical and pharmaceutical companies lead as the most active applicants to the USDA for field testing of transgenic crops.

Lists of approved applications and releases are available at the APHIS Web site (http://www.aphis.usda.gov). The EPA has indicated its intention to classify and thus regulate as "pesticides" all transgenic plants that are disease- or insect-resistant.

TABLE 6.11
Most Active Applicants to U.S. Department of Agriculture (USDA)
for Field Testing of Transgenic Crops

Applicants	% of Applications[a]
Chemical companies	46%
Monsanto (52%)	
Upjohn (Asgrow Seed)	
Dupont	
Sandoz (Northrup, King, and Rogers NK Seed)	
Ciba-Geigy	
Hoechst-Roussel	
Imperial Chemical Industries	
American Cyanamid	
Universities/USDA	17%
USDA Agriculture Research Service	
Cornell University	
North Carolina State University	
University of Kentucky	
University of California	
Michigan State University	
Seed Companies	15%
Pioneer Hi-Bred (51%)	
DeKalb Plant Genetics	
Holden's Foundation Seed	
Petoseed	
Harris Moran	
Biotechnology Companies (biotech only)	13%
Calgene (82%)	
DNA Plant Technology	
Agrigenetics	
Food Companies	5%
Frito-Lay	
Campbell	
Heinz	
Land O'Lakes	
Miscellaneous	4%
Cargill	
Amoco Technology	

[a]Compilation of 549 applications to the USDA from 1987 to 1993.

Source: From J. Rissler, *Perils amidst the Promise: Ecological Risks of Transgenic Crops in a Global Market* (Cambridge, MA: Union of Concerned Scientists, 1993), Figure 1.3, p. 8.

Biotechnology companies have protested the expense and delay engendered by the legal and bureaucratic process of pesticide regulation employed for chemical pesticides. They point out that the same resistance gene introduced by traditional breeding would not be considered a pesticide and would not be regulated. Small companies with a niche product lacking the high-powered regulatory machinery of the large integrated chemical and agricultural companies feel that they are being unfairly excluded from the marketplace. The list of genetically engineered agricultural products currently on the market is short but likely to grow substantially (Table 6.12).

Most of these products promise improved production through pest and herbicide resistance, although agricultural products with genetically altered oil composition are available, and an even more diverse set of products is in the development pipeline. Published Premanufacture EPA Notifications include a number of applications that involve contained fermentation and release into the environment that are directed at production of manufacturing intermediates or products.

Microorganisms are responsible for a large number of commercial products generated through conventional biotechnology. Natural products, or secondary metabolites, are molecules, often chemically complex, that are synthesized by an organism to gain a selective advantage for survival and growth in a particular environment. Some, such as bread mold, have been enshrined in folk medicine by indigenous human societies for treatment of various ailments as ancestors of pharmacological treatment. The most recognizable of these are the antibiotics that have apparently evolved to reduce competition by other organisms for space and nutrients. Their isolation and clinical use as medicines, beginning with penicillin discovered in 1928 by English bacteriologist Sir Alexander Fleming, have been credited, along with great strides in improving public sanitation, for the reduction in deaths due to disease in developed and developing nations since 1900. Microorganisms produce many pharmaceutically useful agents and novel chemical structures. Other materials including food supplements and emulsifying agents for stabilizing fat-containing mixtures are obtained through microbial fermentation. Besides providing scientists with the tools to design new generations of antibiotics to overcome evolving drug resistance, recombinant DNA techniques are used to move the genes responsible for natural product synthesis into organisms more amenable and safer to culture.

TABLE 6.12
Genetically Engineered Agricultural Products on the Market

Product	Company	Enhancement
Herbicide resistant		
Liberty Link® Corn	Bayer Crop Science	Liberty® herbicide resistant
Liberty Link® Canola	Bayer Crop Science	Liberty® herbicide resistant
InVigor® Hybrid Canola	Bayer Crop Science	Liberty® herbicide resistant
Nexera® Canola	Dow AgroSciences	Liberty® herbicide resistant
Roundup Ready® Canola	Monsanto	Roundup® herbicide resistant
Liberty Link® Cotton	Bayer Crop Science	Liberty® herbicide resistant
Roundup Ready® Cotton	Monsanto	Roundup® herbicide resistant
Roundup Ready® Soybeans	Monsanto	Roundup® herbicide resistant
IMI-CORN®	Pioneer Hi-Bred Intl.	Imidazolinone herbicide resistant
IMI™-Canola Seed	Pioneer Hi-Bred Intl.	Imidazolinone herbicide resistant
BXN® Cotton	Calgene, Inc.	Herbicide resistant
DEKALB GR Hybrid Corn	DeKalb Genetics Corp.	Glufosinate herbicide resistant
Pest resistant		
DEKALBt™ Insect-Protected Corn	DeKalb Genetics Corp.	Corn borer resistance
Rogers® Attribute™ Bt Sweet Corn	Syngenta Seeds	Corn borer, rootworm resistant
Maximizer™ Hybrid Corn	Novartis Seeds	Corn borer resistant
YieldGard® Corn Borer	Monsanto	Corn borer and rootworm resistant
Bollgard® Cotton	Monsanto	bollworm, army worm, and loop worm resistant
Freedom II™ Squash	Seminis Vegetable Seeds	Plant virus resistant
Rainbow and Sunuppapayas	Cornell University/Papaya Admin.	Papaya virus resistant
Gray Leaf Spot Resistant	Garst Seed Co.	Disease-resistant corn
Stress resistant		
High pH Tolerant Corn	Garst Seed Co.	Alkaline soil-tolerant
Augmented production		
Optimum® Soybeans	Dupont Agricultural Products	High oleate oil content
Laurical® Rapeseed	Calgene	Special manufacturing food oil
Flavr Runner Peanut	Mycogen	High oleic acid
Naturally Stable Sunflower	Mycogen	No *trans*-fatty acids

Source: Revised from Biotechnology Industry Organization Web site: http://www.bio.org/speeches/pubs/er/agri_products.asp (cited 05/02/05).

Seeking to decrease reliance on petroleum-derived precursor chemicals, a number of industrially useful building blocks such as glycerol and acetone have been produced by microorganisms cultured on low-grade carbohydrate residue from processed agricultural plant waste at competitive cost. Polymeric substances produced by microorganisms such as the food additives xanthan gum and various alginates and the biosynthetic "plastic" polyhydroxybutyric acid (from *Alcaigenes eutrophus*) are available from microorganisms. In 1997, Japanese scientists at research institutes in Tsukuba and Kyoto reportedly engineered a cyanobacterium (*Synechococcus* sp.) to produce up to 10% of its dry weight as poly-hydroxybutyric acid using only light energy and carbon dioxide. Through genetic engineering, the agricultural company Agracetus is constructing cotton plants that will produce fibers of cotton (cellulose) filled with a synthetic polyester fiber—a natural wash-and-wear fiber. Dupont (Wilmington, Delaware) and Genecor International, Inc. (Palo Alto, California) have incorporated genes for bacterial and yeast enzymes in a single organism to make 3G (trimethylene glycol), the monomer base of a new recyclable polyester (3GT). The bottom line remains the economic feasibility of production from renewable biological sources rather than limited petrochemical stores.

The use of transgenic sheep and goats to produce therapeutic products in their milk is now well established. "Pharming" of human proteins can be done on a large, formerly industrial scale, with few effects on the animals. A mature rabbit yields up to 8 liters of milk per year, a sheep 300 liters, a goat 1,000 liters, and a cow 8,000 liters. Thus, small herds of these animals producing at expression levels of 8 to 15 grams of product per liter could be competitive with industrial-scale cell culture of complex proteins. Human anti-thrombin III (ATIII) produced in goats for acquired ATIII deficiency, a blood coagulation proteolytic disorder (Genzyme Transgenics Corp., Framingham, Massachusetts), α1-anti-trypsin, a protease inhibitor produced in sheep for treatment of cystic fibrosis (PPL Therapeutics, Edinburgh, Scotland), and α-glucosidase, a carbohydrate hydrolyzing enzyme, produced in rabbits for the treatment of Glycogen Storage Disease Type II, which results from a deficiency of this enzyme (Pharming Healthcare Products, Leiden, the Netherlands), are all products in various stages of clinical trials. Other proteins are under development, including human serum albumin used as a blood-volume extender for surgery and trauma. Currently, 440 metric tons of the protein extracted from 16 million liters of human blood plasma are required for therapy each year. Transgenic animal production of human proteins would avoid issues of hepatitis, HIV, Creutzfeld-Jacob prion, and other pathogen contamination passed on from human blood supplies, although the potential for transmission of animal diseases across species needs to be addressed. However, governmental action in Europe places the future of this industry in doubt. In March 1998, the Dutch Ministry of Agriculture, partly in reaction to the nuclear transplant cloning controversy, reversed an earlier decision and revoked a Pharming Healthcare Products permit for its nuclear transplant–based transgenic mammal cloning work designed to speed the generation of transgenic animals producing human therapeutic proteins. This judgment effectively bans nuclear transplant cloning in the Netherlands for the foreseeable future.

Bioremediation

As the word's component parts suggest, *bioremediation* means to repair or make right using biological processes. The word is most commonly employed to refer to the removal of manmade contaminants from the environment. In fact, the process is neither new nor restricted to human contamination. Recycling of nutrients, minerals, and chemical building blocks has proceeded for millions of years to make way for new life—and fortunately so, otherwise the earth would be deeply buried under dead bacteria, insect exoskeletons, and leaf litter. With the increasing pressure of burgeoning human populations and the impact of industrialization, humankind has had to find ways to speed up recycling and to clean up concentrations of toxic substances from commercial, governmental, and residential waste and to cope with accidental spills. The difficulty of extracting toxic materials present at the part per million (roughly 1 drop in 13 gallons) level distributed throughout large volumes of earth, water, or air has lead to some innovative treatment strategies. Studies employing naturally occurring microorganisms and plants that use these toxins as energy sources or that concentrate and store heavy metal ions have shown that the desired results can be achieved under some circumstances. Genetic engineering can potentially combine several steps of a metabolic pathway in a single microorganism. A phenotype could be structured to include traits that would not allow the organism to survive outside of the treatment zone, such as requiring the contaminant as a sole carbon source. Fewer types of organisms would then need to be present to degrade many types of contaminants more completely, and the possibility of escape of the altered microbes into the natural bacterial population would be reduced. The widely distributed root systems of some plants can take up toxic metals and, with the help of recombinant DNA technology, can achieve higher specificity and greater concentration of the metals from both land and water. So why aren't these environmental cleanup tools in widespread use? The National Priority List of toxic waste sites contains more than 1,200 confirmed sites with perhaps more than 32,000 sites potentially needing remediation and more being added to the list yearly. In addition, a significant number of the 7 million underground storage tanks in the United States are leaking. This points to an impending crisis of contamination when the tanks reach and pass their designed life spans.

With their capacity to multiply, microorganisms have extended their domain into environments well beyond that which is normally considered conducive to life. From oil reservoirs and ore deposits buried thousands of feet below the earth's surface, to boiling hot springs, to super-heated acid underwater volcanic jets, these single-celled organisms live and reproduce. They thrive on chemicals that we consider toxic waste—sulfuric and phosphoric acid or carcinogen-laced oils and tars—breaking them down into carbon dioxide, water, and mineral elements. The energy and chemical building blocks they extract in this process called *mineralization* are used to make new organisms and to erect barriers against their caustic environment. Soil, water, and air can all be treated by bioremediation. On a familiar level, microorganisms process biological waste in sewage and septic systems all over the world. Microorganisms adapt to utilize the sources of energy and raw materials that are available to them, and those that can't are outcompeted for what food sources remain and are eliminated or reduced to low levels in the microbial population. *Escherichia coli*, the common human intestinal bacterium, contains more than 4,300 protein-encoding genes. In a normal cell growing on glucose as a carbon and energy source, more than 800 enzymes are expressed to catalyze the intertwined cellular reactions we call life. Although *E. coli* are not particularly adept at using environmental contaminants as metabolic fuel, other species such as *Pseudomonas* and *Desulfobacterium* readily adapt, taking advantage of various combinations of metabolic enzymes carried on extrachromosomal DNA plasmids (much like antibiotic resistance factors). Many chemical reactions are carried out—hydrolysis, hydroxylation, methylation and dealkylation, nitro reduction, deamination, ether cleavage, numerous conjugation reactions, and dehalogenation (dechlorination), all of which are needed to break down the toxic chemicals. This last process is particularly useful because the difficult part of many bioremediation efforts is the removal of the highly stable chlorine atoms from industrial environmental pollutants such as DDT as well as those listed in Table 6.13. Often one species of organism carries out one or only a few parts of the degradation pathway, so cooperation between species in the form of a mixed bacterial culture is generally required to degrade resistant compounds completely.

Eastern Europe is a major site for toxic waste cleanup efforts due to lax environmental controls on industry in that region since the Second World War, with an estimated 50,000 contaminated sites in Bulgaria alone. The toxic heavy metal ions copper, zinc, cadmium, arsenic, and even uranium polluting soils can be immobilized by microbial conversion to insoluble forms that leach slowly. Carcinogenic hydrocarbons and long-lived halogenated hydrocarbons are also being targeted for degradation by endogenous bacterial species in several studies. Much of the frontline research is being done by Eastern European scientists in Budapest and Prague, targeting sites such as the Kola Peninsula in Russia (heavy metals, sulfur dioxide), Northern Bohemia (strip mining), and Upper Silesia (industry and poor waste management) (Aldridge 1997).

Biochemical reactions catalyzing the degradation of environmental contaminants can also be exploited to produce salable products on their own. Organisms such as the naturally occurring *Thiobacillus ferrooxidans* and other more thermophilic organisms release economically important metals such as gold and cobalt from their insoluble complexes in ores that are refractory to the normal industrial cyanide leaching process and make recovery of metals from low-grade ores economically feasible. Biocatalytic removal of sulfur from petroleum and coal (desulfuriza-

TABLE 6.13
Examples of Environmental Contaminants

Chlorobenzoic acids	Degradative products of polychlorinated biphenyls (PCBs), herbicides, and plant growth regulators
Chlorinated biphenyls	Electrical transformer and hydraulic fluids, plasticizers, fire retardants
Chlorobenzenes	Industrial and paint solvents, by-products of textile dyeing, fungicides
Chlorophenols	Antifungal agents, wood preservatives, herbicides (2,4-D; 2,4,5-T) derivatives
Phenylamide herbicides	Herbicides
Chlorinated dioxins and furans	Burning of PCBs, manufacturing by-products, hydraulic and heat exchanger fluids

Source: K. H. Baker, and D. S. Herson, eds., *Bioremediation* (New York: McGraw-Hill, 1994), p. 49.

tion) using endogenous bacteria and oxygen to produce cleaner burning fuels is being pursued by companies such as Energies Biosystems Corporation (The Woodlands, Texas) to meet refiners' growing need to decrease sulfur emissions from fossil fuels. Other biorefining technologies being considered include nitrogen and metal ion removal, viscosity reduction, and cracking (hydrocarbon chain length reduction) to make products such as gasoline, processes normally carried out in refineries at high temperature and pressure. The LATA Group (Ochelata, Oklahoma) provides selected water-soluble nutrients to stimulate indigenous beneficial microorganisms in petroleum fields and fuel storage facilities. Damaging bacteria producing the highly corrosive hydrogen sulfide responsible for low petroleum yields and spoilage of petroleum are ecologically outcompeted, reducing the amount of sulfur contaminants in the fuel.

Genetic engineering offers the capability of combining several components of a chemical degradation pathway in a single organism that may also be fitted with biological safeguards beyond natural competition when the foreign substrate is exhausted. A number of types of bioremediation by which different contaminated environmental systems can be treated by altering the method of application of the appropriate microorganisms are described in Table 6.14.

Numerous small companies have sprung up around bioremediation technologies, but the tendency is for them to merge or be taken over by larger organizations that may require their services. International Bioremediation Services (Walsall, United Kingdom), Microbe Masters of InterBio (Baton Rouge, Louisiana), PolyBac (Bethlehem, Pennsylvania), and Oppenheimer Biotechnology (Austin, Texas) are just a few of the survivors with their

TABLE 6.14
Types of Bioremediation

Bioaugmentation	Addition of bacterial cultures to medium
Biofiltration	Microbes immobilized on columns to treat air emissions
Bioreaction	Biodegradation in a container or reactor
Biostimulation	Optimization of microbes already present in a medium
Bioventing	Oxygenating contaminated soils to stimulate microbial growth and activity
Composting	Aerobic thermophilic treatment, addition of bulking agent
Land forming	Solid phase treatment for contaminated soil in place or after removal

Source: Revised from K. H. Baker, and D. S. Herson, eds., *Bioremediation* (New York: McGraw-Hill, 1994), p. 3.

varied technologies. Because each bioremediation situation may require a unique solution, a number of analysts support the formation of bioremediation consortia to broaden their credibility and strengthen their position on legal liability issues. The U.S. Department of Energy (Germantown, Maryland) is charged with environmental cleanup of 2.5 trillion liters of contaminated liquids and some 200 million cubic meters of contaminated solids. The agency, which pursues more traditional methods of environmental restoration, also funds scientific research on alternative methodologies including bioremediation. In 1996, it established BASIC (Bioremediation and Its Social Implications and Concerns), a program designed along the same lines as the Human Genome Project's ELSI (Ethical, Legal, and Social Implications) commission, which is similarly supported by the DOE along with the National Institutes of Health. Like ELSI, this bioremediation organization will have to bridge the often conflicting interests of the scientific researchers, public advocates for community involvement such as the Waste Policy Institute (Washington, D.C.), and government regulators. Preliminary findings from opinion surveys conducted by the governmental agencies Environmental Canada and Industry Canada indicate that the "public expects to be consulted in establishing guidelines or codes of ethics for biotechnology," yet the public lacks specific knowledge about biotechnology and genetic engineering, especially in the area of the environment. Clear explanations will be required to avoid hasty and uninformed public reaction.

Plant-based bioremediation (phytoremediation) is applicable to some aspects of environmental cleanup, although at present it commands a significantly smaller market. The total 1997 U.S. market for phytoremediation was estimated to be $3 to $7 million compared with $200 to $250 million for microbial remediation (D. Glass Associates 1997). The figures were predicted to be more comparable by 2005—$100 to $200 million for plant bioremediation versus $400 to $700 million for microbes. Marshes and estuaries have long been known to cleanse heavy metal ions and excess nutrients in fertilizer runoff from agricultural land. A small number of companies focusing entirely on phytoremediation are experimenting with amaranth, Indian mustard, and sunflowers to absorb toxic or radioactive metal ions from contaminated soil and water with the assistance of the EPA and Phytotech, Inc. (Monmouth, New Jersey). Field trials in Chernobyl, Ukraine, are being run to demonstrate cleaning of the remaining dangerous radioactive cesium and strontium from the atomic reactor accident there. Crossbreeding and genetic engineering are being used to further refine and increase the concentration of these materials by these plants, which can then be safely disposed of or recycled from the harvested plants. PhytoWorks, Inc. (Gladwyne, Pennsylvania) and EarthCare, Inc. (Hanover, New Hampshire) are also concentrating on toxic metal ions such as mercury and on organic chemical contaminants. Phytodegradation of organic chemical contaminants in groundwater is being pursued by Applied Natural Sciences (Hamilton, Ohio), Ecolotree (Iowa City, Iowa), and PhytoKinetics, Inc. (Logan, Utah). PhytoKinetics is also looking at root-based detoxification of organic compounds in soil by grasses and other plants.

Genetic engineering can also enhance phytoremediation. Transgenic tobacco plants making engineered antibodies to atrazine, a widespread herbicide contaminant in soil and water, bind and concentrate the toxin while allowing the plant to survive. In a different approach, Aberdeen University (Scotland, United Kingdom) and Axis Genetics (Cambridge, United Kingdom) plan to transfer the antibody genes to deep-rooted plants such as rape and other *Brassicas,* which would be used as a final cleanup step after most of the atrazine is removed by microbial remediation.

Phytoremediation research is also being carried out at a number of U.S. universities including Cornell University, Iowa State University, Kansas State University, Montana State University, Ohio University, Oklahoma State University, Rutgers University, and the Universities of Georgia, Iowa, Maryland, Missouri, Oklahoma, and Washington. Heffield University and Glasgow University are involved overseas. Various U.S. government agencies sponsor extramural (award grants) or conduct their own laboratory research on phytoremediation, including the Department of Agriculture, the EPA, the Army Corps of Engineers, and the Department of Energy. Some nonprofit organizations such as Argonne National Laboratories, Los Alamos National Laboratories, the Institute of Gas Technology/Gas Research Institute, and the Rothamstead Experimental Station in the United Kingdom also have programs.

Safety

A major impediment to the widespread use of bioremediation to remove environmental contaminants is the question of whether the genetically modified organisms released into the environment will escape control measures and adversely affect the ecosystem. Safety is, of course, a relative term that is defined by tolerability and acceptability limits, which are themselves set according to currently available expertise and, to the chagrin of proponents of environmental uses of genetically modified organisms, increasingly by public reaction. The other part of the evaluation process is risk assessment. This involves a more quantitative measure of the probability of harm occurring, encompassing both the probability of it happening and the amount of damage incurred should it happen. In the case of modified plants or microbes, decisions have primarily been based on safety because it is generally not possible at this time to conduct a straightforward risk assessment on organisms, whether or not they are genetically modified. Chemical toxicology, which has accumulated evidence through long experience, supplies information about classes of chemicals and their effects, the dosages required, and probabilities of adverse effects predicted by validated models as part of the risk assessment–risk management process. With plants, animals, and microbes, although similar organisms and modifications are used to compare with previously analyzed systems, dose-response effects cannot be rigorously determined, whether genetically modified or not. Robust models for living organisms remain to be developed. The particular hazards to be expected may not necessarily be predictable with organisms either. The current compromise for evaluating organisms requires less quantification than chemical risk assessment, but it attempts to identify the hazards of greatest concern. An example of regulatory information required by the European Union is provided in Table 6.15. The difficulty in obtaining this in-depth data in each ecosystem in which the organism is to be used can be readily imagined.

An extensive network of laws regulates environmental remediation in the United States. Phytoremediation is an important component of many treatment and containment efforts. Federal statutes that encompass plant-based technologies include the Resource Conservation and Recovery Act (RCRA); Comprehensive Environmental Response, Compensation, and Recovery Act (CERCLA); a series of Brownfields initiatives; and portions of the Clean Water Act (Flechas and Latady 2003; U.S. EPA 1999).

TABLE 6.15
Regulatory Information for Environmental Release of
Genetically Modified Organisms (European Union)

The Organism
    Organism modification methods
    Construction and implementation of the genetic change (insert or deletion)
    Purity of inserted information and other sequences and functions present
    Similarity in sequence, function, or location of the modified nucleic acid sequences to any known harmful
      sequences
Environmental Impact
    Potential for excessive population increase of organism in environment
    Competitive advantages of modified over unmodified organism in environment
    Anticipated mechanisms and consequences of interactions between modified and unmodified organisms
    Identification and description of nonsimilar organisms that might be affected by genetically modified organism
    Probability of shifts in biological interactions such as host shift after release
    Known or predicted effects on other organisms in the environment such as population changes
    Known or predicted involvement in biogeochemical processes
    Other potentially significant interactions with the environment

EEC Directives:
90/219/EEC Containment
90/220/EEC Release and Marketing
90/679/EEC Health and Safety of Workers Using Genetically Modified Organisms
Adapted from European Union directive 90/220/EEC. Council of European Communities, Official Journal L117/115: 15-27 (1990). In O. Kappeli and L. Auberson, *Trends in Biotechnology* 15 (1997): 342–349.

Source: Revised from European Union Directive 90/220/EEC, 1990.

Ethics

Genetic engineering stirs the primal emotions of right and wrong and fear of the unknown, as discussed in the first part of this chapter and elsewhere in this volume. Yet even if the moral objections to "playing the role of God" are satisfied, further questions remain about the social and personal impact of the knowledge made available by the technology. The following discussion is predicated on general notions of the mores, culture, politics, and economic environment prevalent in the United States. Priorities, accessibility, and the impact of genetic engineering technology are likely to vary in different societies as will the responses of the people in these societies. People's individual feelings of right and wrong may be in distinction to their more homogeneous responses on issues of wider scope such as concern over potential ecological disturbances. Is society ready for the disclosure of hidden susceptibilities and the insecurity of knowing but not truly understanding? The present information glut in areas of much less importance is already hard to handle for many people. Can individuals' right to privacy of their medical information be protected, and how much should be shielded? Can society resist the temptation to interpret probabilities as certainties, denigrating people whose genetic heritage condemns them to an uncertain but threatened future? People's response to currently available biochemical and genetic tests suggests that just because the information is obtainable, it doesn't follow that it will or should be used. Simply banning genetic testing is similarly not a workable option. An unfortunate adjunct of our increasingly litigious society and the availability of genetic testing brings the potential for "wrongful birth" lawsuits by parents and "wrongful life" lawsuits brought by, or on the behalf of, an afflicted child against healthcare providers, if testing was not offered or parties felt that appropriate counseling had not been provided. Leonard Fleck, professor of philosophy and medical ethics at Michigan State's Center for Ethics and Humanities in the Life Sciences, aptly sums up the dilemma: "We used to live in the age of genetic innocence, with no control over our genetic fate or that of our children. Now we live in the age of genetic responsibility."

The National Human Genome Research Institute (NHGRI), a part of the U.S. NIH, established the NHGRI Policy and Legislative database on July 19, 2004, on its Web site (http://www.genome.gov/LegislativeDatabase), which currently focuses on genetic testing and counseling, insurance and employment discrimination, newborn screening, privacy of genetic information and confidentiality, informed consent, and commercialization and patenting.

Bioterrorism

Although not as effective as military weapons, many biological agents have a potential impact sufficient to cause fear and panic in a civilian population. The genomes of many pathological organisms that terrorists might employ have been determined by the Genome Institute for rapid identification of those organisms in suspicious outbreaks of illness. H.R. 3448, the Bioterrorism Preparedness Act of 2002, went into effect on June 12, 2002, charging the FDA with implementation of its measures (http://www.fda.gov/oc/bioterrorism/bioact.html).

Genetic Testing

Approximately 3% of all children are born with a severe disorder generally considered to be genetic in origin. Most genetic diseases manifest early in life, particularly during prenatal development, although there are also a significant number of adult-onset genetic diseases, such as the neurological disease Huntington's chorea. Some examples of currently diagnosable genetic diseases are given in Table 6.16.

Guidelines established by the Institute of Medicine of the National Academy of Sciences for the use of genetic testing depend on the age of the individual being tested, with increased stringency mandated for those considered unable to make informed decisions. The criteria for allowing genetic screening for newborns are that there must be a clear benefit to the newborn, confirmatory

TABLE 6.16
Examples of Prenatally Diagnosable Genetic Diseases

Disease	Incidence	Symptoms
Phenylketonuria	1:10,000	Mental retardation
Hemoglobinopathies		Anemia
Sickle cell	common in Africa and Mediterranean	Anemia, ischemic crises
α-thalassemia	common in Mediterranean and Asia	Anemia, ischemic crises
α1-antitrypsin deficiency	1:8,000	Emphysema (lung), liver disease
Hemophilia A	1:10,000 males	Bleeding disorders
Tay-Sach's disease	1:300,000 general population 1:3,000 Ashkenazi Jews	Mental retardation
Gaucher's disease	Rare in general population 1:600 Ashkenazi Jews	Anemia, enlarged spleen
Glucose-6-phosphate	Variable, many mutations	Anemia, dehydrogenase deficiency
Type II hyperlipidemia	1:2,500	Atherosclerosis, coronary heart disease
Familial hypercholesterolemia	1:500	Atherosclerosis, coronary heart disease
Duchenne muscular dystrophy	1:3,000	Muscle wasting
Cystic fibrosis	1:2,000–3,000	Lung failure, digestive malabsorption
Neurofibromatosis	1: 3,500	Nervous system tumors
Adult polycystic disease	1:5,000	Kidney failure
Huntington's disease	1:10,000–12,000	Uncontrolled movement — nervous system degeneration
Spinocerebellar ataxias	1:25,000–50,000	Movement disorders — nervous system degeneration

Source: Revised from C. T. Caskey, in *Code of Codes,* D. J. Kevles, and L. Hood, eds. (Cambridge, MA: Harvard University Press, 1992), Tables 3-6, pp. 119, 122, 127, 133.

tests must be available, and both treatment and follow-up are available for the affected individuals. This specifically forbids determining whether a newborn carries a genetic trait for a disease as opposed to determining whether the child has a predisposition to the disease. If a high risk of carrying a disease gene is suspected in a family, then testing of the parents is recommended.

In general, testing of children is indicated only if there is a curative or preventative treatment that must applied early in life to be effective. Carrier status testing or testing for incurable or late-onset diseases (unless preventable by early treatment) is to be deferred to adulthood when the individual is presumably capable of informed decisions. The reasoning includes the concept that unrestrained childhood testing infringes on that individual's confidentiality rights (normally provided to adults). Knowledge of a child's genetic status risks stigmatizing his or her upbringing and relationship to family members and raises life and health insurance issues. Finally, testing a child also intrudes on his or her future right to make the decision for testing as an adult. Another reason for this is that the majority of eligible adults (depending on the disease), in fact, choose *not* to be tested. One-third of individuals at risk for Huntington's disease did not plan to make use of genetic testing (Adam et al. 1993). Although 75% of partners of at-risk individuals were in favor of genetic testing, only 29% of the at-risk individuals were supportive. Thirty percent of pregnant at-risk individuals requested the testing, and only 18% actually had it performed (Elvers-Krebooms 1990). Nancy Wexler, a Ph.D. clinical psychologist at the California-based Hereditary Disease Foundation, herself at risk for Huntington's disease, remarked in 1984, "It's not a good test if you can't offer a treatment." Eighteen percent of parents with a child afflicted with phenylketonuria were willing to undergo prenatal diagnosis in subsequent pregnancies (Barwell and Pollitt 1987). Eighty-one percent of women at increased risk for fragile X syndrome (Moorish and Abuelo 1988) and 82% of carriers of hemophilia (Lajos and Czeizel 1987) would choose prenatal testing under the same circumstances.

Arguments in support of genetic testing of children can be made on the basis of a parent's right to know and resolution of parental uncertainty as well as on providing lead time for psychological adjustment. There are also families in which siblings have already have been tested. Proponents of childhood testing also point out that based on genetic statistics, between 33% and 50% of those tested will be reassured by the results that they don't have the potential for the disease. Others reported "survivors' guilt" as well as the realization that as a consequence of not carrying the genetic trait, they will be responsible for the care of afflicted siblings.

Prenatal testing for catastrophic fetal disorders by looking for biochemical expression of the disease has been mandated by many states for decades (Table 6.17).

In contrast to genetic analysis, there has been little criticism of this type of testing. These tests more closely measure the clinical impact of the disease rather than a risk or probability. Advances in genetic technologies made possible by recombinant DNA techniques are now capable of detecting many more genetic abnormalities with higher sensitivity than the biochemical tests. In addition, biochemical defects may show up later. A list of some of the techniques now in use along with their pros and cons is provided in Table 6.18.

So why isn't all testing done by genetic means? The reason is that the complexity of factors involved in producing most disease states makes it difficult to predict exactly who will develop the disease and when. The tests determine only the average probability of developing the disease at some time during a lifetime. Biochemical tests based on the abnormality only identify those who have clinical expression of the condition at the time of testing.

TABLE 6.17
U.S. Screening of Newborns for Genetic Disorders: 2004

Disorder	No. States Promoting Screening[a]
Phenylketonuria	52
Congenital hypothyroidism	52
Galactosemia	52
Hemoglobinopathy	50
Adrenal hyperplasia	40
Maple syrup urine disorder	37
Biotinidase deficiency	34
Homocysteinuria	33
Tyrosinemia	20
Cystic fibrosis	7
Toxoplasmosis	3

[a]Including District of Columbia, Puerto Rico, and U.S. Virgin Islands.

Source: National Newborn Screening and Genetics Resource Center, U.S. National Screening States Report, November 15, 2004.

TABLE 6.18
Genetic Testing Techniques: Scanning for Mutations

Nucleotide sequencing—Boosted by technology developed for the Human Genome Project, highly parallel capillary sequencing is rapid and inexpensive. Candidate gene regions are sequenced.

Chip technology—Microfabrication techniques allow thousands of DNA sequences to be tested on a single chip. Candidate gene regions are probed.

Protein truncation—This is the only test that measures whether a mutation in a candidate gene is expressed because it tests for changes in the protein product of that gene. The mutation must cause the protein sequence to be shorter than normal to be detected. Rarely, a protein will be longer than it should be if the mutation causes the protein synthesis machinery to go past the normal stop signal.

PCR mRNA amplification—Polymerase chain reaction can be used to amplify and sequence cDNA made from the mRNA of a test tissue to look for mutations that are expressed. This method is almost as direct as the protein truncation method but can detect single nucleotide changes.

Fluorescent In Situ Hybridization (FISH)—Deletions or translocation of parts of chromosomes to another chromosome are common changes in certain kinds of cancer that are followed by fluorescent tags directed to specific DNA sequences. Genes can be fragmented, controlled by inappropriate promoters, or shut off by chromosomal regulators. Duplications of parts or whole chromosomes such as chromosome 21 in Down syndrome are also detected by this method.

Source: Revised from C. Eng, and J. Vijg, *Nature Biotechnology* 15 (1997): 424–426.

They cannot predict those who will develop the disease later in life. One in twelve African Americans is a carrier of the sickle cell trait, a hemoglobin variant that affects the red cell response to oxygen. If both parents carry that recessive trait, their child has a 1 in 4 chance of having sickle cell anemia. Yet with prophylactic penicillin treatment to prevent infections, the observed incidence of sickle cell anemia is actually 1 in 600. Thus, having the gene variant does not guarantee having the disease. Other tests for later-onset genetic diseases are performed on adults suspected to be at risk because of a family history of a disorder after they have decided that they want to know whether they have the disorder or are a carrier (Table 6.19).

Even though there are few individuals with identified genetic diseases (see Table 6.17 for incidence), many people are involved in decisions about genetic testing. Thirty-seven percent of respondents to a survey by the National Opinion Research Center in 1990 ("Genes, Social Science Survey" 1990) report having an immediate family member who has had or who is at risk for, or who is a carrier of a genetic disease. This surprising percentage is likely to be an overestimate because in identifying the dis-

TABLE 6.19
Examples of Late-Onset Genetic Disorders

Disease	Incidence	Pathology
Monogenic		
Huntington's chorea	1: 100,000	Uncontrolled movement
Hemochromatosis	1:500 (Caucasian)	Iron storage
Familial hyper-cholesterolemia	1:500	Atherosclerosis
Polycystic kidney disease	1:500–1:5,000	Renal failure
Inherited susceptability to cancers	1:200	p53, breast cancers
Multigenic		
Congestive heart disease	Several major susceptibility genes	
Hypertension	Several genes regulating sensitivity	
Cancers of complex origin	Not inherited—somatic changes in tissue	
Diabetes	Numerous genes	
Rheumatoid arthritis	HLA antigens and others	
Psychiatric disorders	Schizophrenia, manic depressive disorder, panic disorders, Tourette's syndrome, some alcoholism	
Schizophrenia	Genes on several chromosomes	
Bipolar disorder	Genes on several chromosomes	

order most people in the survey focused on physical deformities or on mental retardation, which in many cases actually may be environmental or traumatic and not genetic in origin. Much the same as surveys reporting a high level of scientific illiteracy, 85% of the respondents to this survey claimed to have read or heard little or nothing about genetic screening. Nevertheless, they express strong opinions about screening and are wary of the potential for misuse. A March of Dimes survey also found that 68% of respondents knew little or nothing about genetic testing, and 87% were similarly ill-informed about gene therapy (March of Dimes 1992).

Employment and Insurance

Employers in increasing numbers are using genetic testing in support of hiring decisions as an adjunct to a battery of psychological and standardized tests. The objective is to reduce the risk that a new employee will be transient, absentee, or disruptive or will increase costs of Workmen's Compensation coverage or medical insurance. Identifying and excluding people with a propensity to allergies or a genetic susceptibility to cancer from working in environments with possible exposure to levels of agents that have little effect on the general population would be another use for such information. Opponents of this type of testing demand that instead the employer should be responsible for reducing the environmental risk to permit even highly sensitive individuals to work there. Such uses of genetic testing argue that a genetic predisposition can be equated with a preexisting condition. The logic of this reasoning is flawed because for most genetic disorders the genetic marker reflects only a probability that a person will develop a disease, not that he or she has the disease. Opposition to the use of genetic testing for setting selection criteria that are beyond an individual's control, such as race or genetic background, comes from the perception of many that such practices are unfair and discriminatory.

Fewer than half of the states have addressed the employment issue, and the federal government only began to consider legislation on the privacy of genetic information in 1996. Fourteen states (Arizona, Florida, Illinois, Iowa, Louisiana, New Hampshire, New Jersey, New York, North Carolina, Oklahoma, Oregon, Rhode Island, Texas, and Wisconsin) restrict employer access to and use of genetic information. Some are more comprehensive than others in their protection. Florida and Louisiana only cover sickle cell disease, and Oklahoma has only set up a task force to make nonbinding recommendations to the state legislature. The other eleven states with legislation forbid discharging or refusing to hire an individual on the basis of genetic information. Seven states (Arizona, Iowa, New York, Oregon, Rhode Island, Texas, and Wisconsin) forbid employer access to results of genetic testing without an individual's consent, and some allow no access at all. Iowa, New Hampshire, New Jersey, New York, Oregon, Rhode Island, and Wisconsin forbid employers from requiring genetic testing as a condition of hiring or continued employment. New Hampshire, New Jersey, New York, and Wisconsin permit occupation-related disorders susceptibility testing if requested by an individual seeking employment who might be worried about his or her sensitivity to environmental conditions.

People are overwhelmingly opposed to using genetic testing in employment decisions: 85% (U.S. General Social Survey 1990) and 80% (ABC News Poll, May 1990). Seventy-five percent supported the principle that the tested individual should retain sole control over access to his or her genetic testing information (U.S. General Social Survey 1990). The sentiment is mirrored in Germany with 75% opposed to the use of genetic testing in employment screening, and 23% favoring prohibiting genetic testing altogether (Hennen, Petermann, and Schmitt 1990). However, 26% of the respondents in the German survey thought that people have a duty to have their genetic makeup tested.

A similar campaign is being waged to prevent or restrict the use of genetic information by the health and life insurance industries to determine the cost of covering individuals. In actuarial analysis, risk of a particular outcome is calculated for various categories of people who have been grouped by similarities in information available to the company, age, high blood pressure, cholesterol, family history, smoking history, and so on. The companies would like to use genetic testing to further stratify their clients to improve their ability to predict risk (and some people believe simply to exclude people with the potential to develop expensive diseases). Although this would protect the company against losing money on payments for exceptionally high-risk individuals, charging premiums adjusted for the individual risk would make it even more difficult for people with chronic diseases to find and retain insurance coverage at an affordable cost. Denial of insurance coverage is not an exceptional event. In 1987, health maintenance organizations denied membership to 24% of individual applicants (not group applicants, who are usually covered through their place of employment or other organization).

The Health Insurance Portability and Accountability Act of 1996 (HIPAA) prohibits the use by employer-based or commercial health insurers of either genetic test results or family history to deny coverage or charge high rates to individuals based on a pre-existing condition unless the disease is already clinically active. This protects some 150 million Americans insured through group plans primarily by their places of employment. Thirteen million Americans with individual healthcare insurance policies enjoy no such refuge. The National Human Genome Research Institute reported that most states have enacted legislation to conform with HIPAA. Thirty-seven states also enacted legislation on genetic discrimination in health insurance, and 24 states enacted legislation on genetic discrimination in employment.

Fear of having their genetic information used against them looms large for many people. An NIH study found that 32% of women approached to take part in a genetic breast cancer detection trial did not participate, mostly due to fear of discrimination and lost privacy. One woman, whose family includes nine breast cancer cases in three generations, is waiting until legislation guaranteeing nondiscrimination is on the books: "My doctor said to me, 'If you get this test done, your daughter is not going to get insurance.'"

How can the needs of people to have affordable healthcare regardless of their genetic and economic status be provided for in the absence of an effective state or national healthcare program? Assuming that this is approached through insurance plans and that the companies are entitled to use all available information to match each insuree's premium payment to their individual risk, there are several possibilities. One compromise would be for all of the insured to pay more to subsidize those more likely to require benefits. This would yield essentially the present situation when specific genetic information is not known. Another would be to provide an adequate basic level of coverage for everyone without the use of genetic information with the opportunity for additional coverage using all available information including genetic testing to set the rates. Alexander Tabarrok, assistant professor of economics at Ball State University, suggested yet another option: People, before opting for genetic testing, could purchase insurance against finding that they carry a disease-causing gene. This insurance, if a person tested positive for a genetic disease marker, would cover the higher premiums charged by an insurance company, the higher healthcare costs, or both.

The rapidly expanding capabilities and use of genetic testing in personal health and employment decisions have prompted the drafting of the legal basis to control abuses of the new information. In February 2000, President Bill Clinton signed an executive order prohibiting federal government agencies from obtaining genetic information from employees or job applicants and from using genetic information in hiring and promotion decisions. Three bills were introduced in the 107th Congress to extend this sort of protection to the private sector: HR 602 and S. 318 Genetic Nondiscrimination in Health Insurance and Employment Act and S. 383 Genetic Information Nondiscrimination in Health Insurance Act of 2001.

Some people feel that blocking access to genetic information will be futile and that regulation would be more effective if targeted against the *use* of the information. Hastily proposed in early 1997, none of this legislation was enacted for a number of years because of concern over the potential limitation on access that could include genetic information such as that collected under informed consent in clinical trials. Industry organizations such as the Pharmaceutical Manufacturer's Association feel that such blanket denial of access will severely restrict the use of anonymous genetic information to devise new medical therapies.

The direct marketing of genetic testing to the public is now a reality. Some commercial laboratories such as Myriad Genetic Laboratories, Inc. (Salt Lake City, Utah) and Oncormed, Inc. (Gaithersburg, Maryland) who offer breast cancer gene screening do so only to a group of select clients. They also provide instruction to physicians about how to screen prospective patients and provide educational materials for medical professionals and consumers. In addition, these companies assist physicians in locating genetic counseling for their high-risk patients. There are fears of unregulated marketing to the general public without providing adequate interpretation and counseling to make use of the results, a situation characterized by Glenn McGee (Center for Bioethics, University of Pennsylvania) as "drive-through genetic testing." As part of the Human Genome Project, the ELSI Task Force on Genetic Testing issued a series of recommendations in May 1997 that included the following: 1) institutional review board approval of predictive genetic tests; 2) outside review of clinical utility and validation data provided by test developers; 3) enhanced genetic training of healthcare professionals including nurses, social workers, and public health workers whose patients have diseases with substantial inherited components; 4) development of stringent criteria of predictive genetic tests; 5) evidence of competency to interpret genetic tests and provide counseling for hospitals or managed care organizations before ordering genetic tests; and 6) improved transmission of new information on rare genetic diseases to physicians. These recommendations remain to be completely codified.

Genetic testing may soon be found alongside the blood glucose and pregnancy home testing kits. Gamera Bioscience Corp. (Medford, Massachusetts) is developing the LabCD System, a compact disk player–sized unit that will perform several genetic tests simultaneously. While initially targeting test manufacturers, the company has an eye on physician's offices and even the home, Gamera Chief Executive Officer Alec Mian has indicated. How will enforcement of the laudable safeguards of the ELSI Task Force on Genetic Testing against misuse and abuse of genetic information be assured with the availability of "on-demand" testing?

DNA Forensics

Several commercial companies have specialized in performing forensic DNA analysis, which is beyond the scope of most forensic laboratories, although the FBI maintains its own forensic DNA capability. Cellmark Diagnostics, a subsidiary of Imperial Chemical Industries, with labs opening in Germantown, Maryland, in 1987, pioneered the DNA "fingerprinting"[SM] technology based on restriction fragment length polymorphism (RFLP) analysis. Lifecodes Corporation (Valhalla, New York) also used an RFLP technology while Forensic Science Associates (Richmond, California) has pursued polymerase chain reaction amplification and marker alleles of known genes such as human leukocyte antigen (HLA) DQα–1 locus. Gennan Corporation (Akron, Ohio) and Genescreen (Dallas, Texas) have also been identified as commercial forensic DNA analysis laboratories by the U.S. Office of Technology Assessment. The completion of the human genome sequence has revolutionized the area of DNA identification by establishing norms for the extent of differences between individuals. Through January 1990, 185 court cases were reported using DNA typing evidence in the prosecution or defense (Table 6.20). Often these cases are crimes of violence where there

TABLE 6.20
Reported Uses of DNA Typing and DNA Database Legislation (2003)

Offense	No. States[a]
Sex crimes	50
Murder	50
All violent felons	47
Burglary	46
Drug crimes	38
All felons	34
Juveniles	31
Some misdemeanors	26
Arrestees	3
Jailed offenders	48
Community corrections	48
Retroactive jail and prison	36
Retroactive probation and parole	22

[a]Does not include the District of Columbia or Puerto Rico.

Source: Revised from DNARESOURCE.COM. Available: http://www.dnaresource.com/bill_tracking_list.htm (cited 07/01/05).

is no living witness, thus rendering the DNA evidence crucial in obtaining a conviction.

The tendency toward collection of large databases of information operates in law enforcement perhaps with more energy than in any area other than the military. By January 2003, the United States held more than 1 million DNA profiles from convicted offenders and more than 50,000 profiles from crime scenes; 40,000 matches of suspects have been made to a crime scene (http://www.ornl.gov/sci/techresources/Human_Genome/elsi/forensics.shtml). Law enforcement officials cite the high rate of recidivism among convicted violent offenders as justification for the maintenance of databases to aid in identification and swift prosecution. In 1989, the Bureau of Justice Statistics reported that 62.5% (59.6% of the violent criminals) of prisoners released in 1983 had been rearrested within 3 years, with 41.4% being returned to prison. Among the violent offenders, rapists were 10.5 times more likely than other released offenders to be rearrested for rape; murderers about 5 times more likely than other offenders to be rearrested for homicide. Although these statistics would argue powerfully for databases to protect society, it is noteworthy that in the same study only 6.6% of the released rapists were rearrested for rape and 7.7% of the killers were rearrested for murder (Bureau of Justice Statistics 1989). The civil rights of convicted criminals, particularly felons, are legally curtailed compared with those of the average citizen. The Privacy Act of 1974 (5 U.S.C. 552a) governing data collection and access to information about most people in federal databases specifically exempts criminal justice agency record systems from many provisions (5 U.S.C. 552a(b)(7), (c)(3), and (j)(2)). States control the privacy of nonfederal criminal history databases, ranging from complete public accessibility (Florida) to sealed records (Massachusetts). Pressed by rising crime statistics, state legislatures passed legislation to require or propose the collection of DNA samples and test results from certain convicted offenders, particularly those who crimes of a sexual nature (see Table 6.20) and (http://www.dnaresource.com/bill_tracking_list.htm). The creation of a national DNA database could aid the criminal justice system and the military, but it would be essential in many people's minds to restrict the use of that information. The U.S. military maintains the Armed Forces Repository, a DNA database to aid in identifying remains (http://www.afip.org/Departments/oafme/dna/afrssir/). As of March 1, 2006, over 4.6 million specimens had been collected from personnel in all branches of the military and the Coast Guard. While there is a stated policy governing sample use, critics still question how long the DNA will be held and under what conditions information from that DNA could be used.

Human Cloning

The announcement on February 22, 1997, of the successful differentiated cell nuclear transfer, dubbed "cloning," of Dolly, a Finn Dorset lamb (Wilmut 1997), and shortly thereafter that of non-genetically identical rhesus monkeys using embryonic nuclear transfer to enucleated mature oocytes at the Oregon Regional Primate Research Center was followed by several similar experiments with other animals. These achievements elicited an immediate call for legislation to control, if not outright ban, the application of similar technology to humans. On February 27, 1997, S. 368, restricting the use of federal funding for research related to cloning, was introduced into the Senate. This was swiftly followed on March 5, 1997, by the introduction of H.R. 922 into the House of Representatives with a similar intent. Significantly, this was supplemented with H.R. 923, which would make it a civil crime for a person to conduct cloning of a human. As introduced, these bills contained little more than the initial statement of purpose. The intention was that details were to be added as the bills made their way through the legislative process. People trying to keep in sight the broader vision for society injected a note of caution about blanket bans on cloning research. The chairman of the genetics subcommittee of the National Bioethics Advisory Commission, Thomas H. Murray, Ph.D. (director of the Center for Biomedical Ethics, Case Western Reserve University School of Medicine, Cleveland, OH), remarked at the March 5 House Science Subcommittee meeting, "It is essential that we keep in view the possible scientific, and ultimately the human, benefits of research on animal cloning, benefits best described by other witnesses before you today. It is important that our public policy response to research on the cloning of animals not be swept along by our concern to prevent what we will judge to be the ethical dangers of human cloning."

H.R. 922—The Human Cloning Research Prohibition Act would prohibit the "expenditure of federal funds to conduct or support research on the cloning of humans." Introduced by Vernon J. Ehlers (R-MI) on March 5, 1997, it was referred to the Committee on Commerce and the Committee on Science.

H.R. 923—The Human Cloning Prohibition Act would make it "unlawful for any person to use a human somatic cell for the process of producing a human clone." It imposes a civil penalty of up to $5,000. Introduced by Ehlers on March 5, 1997, it was referred to the Committee on Commerce.

S. 368 aims "to prohibit the use of federal funds for human cloning research." This bill defines cloning as "the replication of a human individual by the taking of a cell with genetic material and the cultivation of the cell through the egg, embryo, fetal, and newborn stages into a new human individual." Introduced on February 27, 1997, by Kit Bond (R-MO) and John Ashcroft, it was referred to the Committee on Labor and Human Resources.

Before the passage of legislation, a voluntary moratorium on human cloning was generally agreed on by scientists, medical professionals, legislators, and industry. Lacking the force of law, there is historical precedence for the effectiveness of such voluntary restrictions with the initial use of recombinant DNA technology in the early 1970s. At that time, potentially risky experiments were delayed for several years until detailed guidelines could be developed and safety issues had been resolved. Analogous guidelines are expected to evolve. News opinion polls conducted soon after the cloning announcements showed a significant amount of concern: 56% (America Online) and 93% (Times—CNN) of Americans opposed human cloning.

At the request of President Clinton, the National Bioethics Advisory Commission called for immediate action on the following:

"1) A continuation of the current moratorium on the use of federal funding in support of any attempt to create a child by somatic cell nuclear transfer.

2) An immediate request to all firms, clinicians, investigators, and professional societies in the private and non-federally funded sectors to comply voluntarily with the intent of the federal moratorium. Professional and scientific societies should make clear that any attempt to create a child by somatic cell nuclear transfer and implantation into a woman's body would be at this time an irresponsible, unethical, and unprofessional act." Full report National Bioethics Advisory Committee, "Ethical Issues in Human

Stem Cell Research," 1999. Available http://www.georgetown. edu/research/nrcbl/nbac/pubs.html.

To address the general public concern about human cloning, scientific societies such as the Federation of American Societies for Experimental Biology (FASEB) adopted specific resolutions supporting the moratorium and called upon scientists to provide input to ensure that imprecise or misused technical language in anti–human cloning legislation did not compromise vital biomedical research. For instance, *cloning* is a general term that includes general recombinant DNA handling as well as somatic cell nuclear transfer. Research involving "cloning" the human insulin receptor and expressing it in a cultured cell line to learn more about how insulin works and what might be going wrong in diabetes could be prohibited by sloppily written legislation. The FASEB Public Affairs Executive Committee adopted the following resolution on September 10, 1997:

> Resolved: The Federation of American Societies for Experimental Biology (FASEB) adopts a voluntary five-year moratorium on cloning human beings, where "cloning human beings" is defined as the duplication of an existing or previously existing human being by transferring the nucleus of a differentiated, somatic cell into an enucleated human oocyte, and implanting the resulting product for intrauterine gestation and subsequent birth.

Although seeming needlessly complex, the detail in the legislation is essential; it avoids affecting cloning of human genes for medical research as well as the scenario of a woman being prosecuted for "cloning" by bearing monozygotic (maternal or identical) twins. In fact, mammalian cloning has been widespread for a number of years with the cloning and implantation of prize-winning cattle embryos. It is not entirely clear to some people why cloning people presents such a problem because, in effect, with identical twins there are many "clones" already around. Society deals with them as individual persons with little remark. They are identical genetically which, as has been pointed out in a previous section, only addresses potentiality, not actuality. A number of scientists and the general public, particularly in the United States, now wonder whether our leaders were stampeded into poorly thought out legislative action. The broadly written Human Cloning Prohibition Act, S. 1061, which would have made human somatic cell nuclear transfer a criminal offense, punishable by fines and a 10-year prison sentence, had been hurriedly introduced. During the February 11, 1998, debate on S. 1061, Senator Connie Mack, himself with cancer, whose parents and brother died of cancer, and whose wife and daughter were similarly diagnosed, offered this impassioned plea:

> I appeal to you, don't get drawn into this debate that we should pass this legislation because we want to stand up and make a statement that we are against cloning. We are all against human cloning. What I am asking you to do is to vote no on cloture so we will have an opportunity to hear from those patient groups that want to represent people like myself, represent families that have been affected like my family has been affected.

The cloture vote was to cut off debate, forcing an immediate vote on the bill without the discussion of a full hearing to present the information needed to make an informed decision on this highly technical yet highly emotional issue. Senator Robert C. Byrd (D-WV) offered the following caution:

> Who can say with any comfort what the impact may be on important research aimed at dread diseases? Doesn't important and potentially far-reaching legislation such as this at least warrant hearings before we proceed? This legislation could have unintended and detrimental consequences.

By a count of 42–54, short of the 60 votes needed to end debate, the Senate in effect decided to continue the discussion. A substitute Human Cloning Prohibition Bill, S. 1062, jointly sponsored by Senators Diane Feinstein (D-CA) and Edward Kennedy (D-MA), which bans the implantation of the product of nuclear transplantation in a woman's womb, but not the nuclear transplantation technology itself, was put into consideration. Although being explicit about restricting human somatic cell nuclear transplantation for the purpose of creating a child, specific protections for nuclear transfer in nonhuman animals, general DNA cloning, and vital medical research are provided.

BIO, the Biotechnology Industry Organization, representing more than 745 biotechnology companies, academic institutions, and state biotechnology centers in 46 states and more than 25 countries, agrees that cloning of a human being is "plainly inappropriate." Carl B. Feldbaum, president of BIO, issued the following statement (in part) (PR Newswire 1998):

> BIO continues to support President Clinton's and the National Bioethics Advisory Commission's (NBAC) moratorium on the cloning of a human being. Cloning a human being poses major ethical and moral questions, as well as deeply troubling medical safety issues. . . . the public has come to learn of the benefits of cloning cells, genes and tissues, techniques which have been ongoing for 20 years. While these techniques do not lead to the cloning of a human being, they do have promise, for example, to enable us to regenerate spinal cord tissue for accident victims and skin for burn victims. It is essential that any restrictive legislation which may be adopted by Congress or state legislatures on the cloning of a human being, recognize the biomedical benefits of these existing cloning procedures and protect and encourage them.

The search continues for a legislative solution that balances a spectrum of ethical and political viewpoints with both the freedom to avoid blocking medical advances and the recognition of the global spread of the technology. Many bills are proposed, but most are not enacted, so it is confusing to determine where the law of the land stands. In addition, the development of stem cell technology further complicated issues and required modification of legislation. Some examples of recent legislation are the following:

H.R. 534—The Human Cloning Prohibition Act of 2003 (D. Weldon, R-FL) prohibits both reproductive and therapeutic cloning and provides a criminal penalty of up to 10 years for violation. Passed 241–155. A substitute measure, H.R. 801 (J. Greenwood, R-PA), that would have prohibited only reproductive cloning failed 174–231.

H.R. 916—The Human Cloning Research Prohibition Act (C. Stearns, R-FL) would prohibit the use of federal funds to conduct or support research on human cloning. It would also express the Sense of Congress that other countries should legislate similarly.

H.R. 938—The Human Cloning Prevention Act of 2003 (R. Paul, R-TX) would prohibit any federal agency from supporting any entity that "within the past year has engaged in human cloning," with human cloning defined as including somatic cell transfer to derive stem cells.

H.R. 1357—The Human Cloning Prohibition Act of 2005 (D. Weldon, R-FL) would make it a criminal act with a penalty of up to 10 years and/or a fine of $1 million, or twice any monetary gain over $1 million, for any individual or entity public or private to knowingly participate in human cloning.

S. 245—The Human Cloning Prohibition Act of 2003 (S. Brownback, R-KS) is similar to H.R. 234 and would prohibit both reproductive and therapeutic cloning.

S. 303—The Human Cloning Ban and Stem Cell Research Protection Act of 2003 (A. Specter, R-PA) would prohibit reproductive cloning but specifically permits therapeutic cloning if carried out within certain ethical guidelines.

The international reaction to Dolly was similar to that in the United States, swift and severe. Great Britain initially banned human cloning outright, while a number of European countries including Germany and Italy initially called for either worldwide bans or instituted temporary moratoria on human cloning experiments. In 2001, Great Britain became the first country to legalize research cloning. One license in August 2004 was granted to a research team to create insulin-producing cells for transplant into diabetics. A second was granted in February 2005 to derive stem cells from patients with motor neuron disease (Lou Gehrig's disease) to induce them to form neurons and to study how they differ from normal motor neurons (Wagner 2005). According to the Center for Genetics in Society (http://www.genetics-and-society.org/), of 192 countries, 23% ban reproductive cloning, and 77% have not taken action. Six percent allow cloning for research purposes, and 16% ban cloning for research.

Stem Cells

The furor over reproductive cloning, popularly known as human cloning, was still in full swing when the derivation of human pluripotent stem cells from human blastocysts (Thomson et al. 1998) and germ cells from cultured primordial human germ cells (Shamblott et al. 1998) was announced. In some people's eyes, this immediately inserted abortion and the issue of viability of embryos used to derive embryonic stem cells into concerns over the technology. Although not the realm of genetic engineering, this form of biotechnology is linked in the public mind with nuclear cloning (reproductive cloning) and with the potential to introduce genetic modifications. The Clinton administration issued temporary guidelines governing the use of embryonic stem cells on August 23, 2000, and postponed review of grant applications until the Bush administration could assess the situation. President George W. Bush issued his decision on funding of stem cell research on August 9, 2001, that restricted funding to research on stem cell lines established before August 9, 2001.

Numerous bills to regulate stem cell research were offered. Because of the potential medical benefits and intensive lobbying by researchers, biotechnologists, and patients, the legislation has been less restrictive. In addition, the regulations only apply to work supported by federal funds, not to the private sector. The following legislation is under consideration in both Houses of Congress:

H.R. 2059 (J. McDermott, D-WA) and S. 723—The Stem Cell Research Act of 2001 (A. Specter, R-PA) would provide for human embryonic stem cell generation and research.

H.R. 2096 (C. H. Smith, R-NJ) and S. 1349—The Responsible Stem Cell Research Act of 2001 (J. Ensign, R-NV) would establish a stem cell donor bank derived from adult tissue, placentas, and umbilical cord blood. No embryos are involved. These stem cells are not embryonic and not fully pluripotent. They are capable of differentiating into fewer types of cells than embryonic stem cells.

H. Con. Res. 17 (C. Maloney, D-NY) is a congressional resolution expressing the Sense of Congress on pluripotent stem cell research.

H.R. 2747 (D. L. DeGette, D-CO) would convert the Clinton guidelines of 2001 on research with pluripotent stem cells into law.

H.R. 2838 (J. Millender-McDonald, D-CA) would require NIH to conduct human embryonic stem cell research and would repeal the Human Embryo Research Ban contained within the Labor, HHS, and Education Appropriations Act.

The failure of the prohibition of alcohol manufacture and sales in the early 20th century is a recent demonstration that regulation is much more effective at controlling citizen behavior than outright banning of activities. A sign of this realization is introduction of legislation to police rather than ban stem cell research.

H.R. 2863 (J. McDermott, D-WA) would establish an FDA Advisory Committee to recommend policy on human embryonic stem cell research and therapeutic cloning.

H.R. 4011 (C. Maloney, D-NY) would establish a Stem Cell Research Board within the legislative branch to evaluate the effects of President Bush's August 9, 2001, policy decision on stem cells.

Summary

Genetic engineering is poised to make its greatest impact over the next few decades. It is hard to judge the balance between the relative good and bad potential of the technology. To this point, we have only been able to catch a glimmer of the potential advances to be made with this newfound power to see into and to influence nature on a scale never before imagined. This power is frightening to many people. It has forced us to think both about our place in the world ecological web and about our relationships to one another. Powerful tools to improve our understanding of the chemistry of life and to probe what goes wrong in disease are the major accomplishments of the first 30 years of genetic engineering. New medicines have come from this, certainly, although full eradication of a genetic disease is still in the future. Both are positive contributions. Less understood and therefore feared by many are potential repercussions in the environment from the wide-scale release of genetically modified plants, animals, and microbes in industrial, agricultural, or environmental remediation applications of genetic engineering. Acceptable control measures remain to be developed, as the technology is available. Banning it will only force applications underground and generate suspicion. Negative repercussions of genetic engineering, at least until society figures out how it wants to regulate it, are most notable in the genetic testing and forensic realms. Controlling the application of the information obtained from the testing and from databases has the potential to determine whether the impact of genetic engineering will be good or bad. These issues have also surfaced in discussions about reproductive versus therapeutic cloning and stem cell research and potential therapies.

Genetic engineering technology applied to human health challenges some of our deepest feelings about our private selves and makes us reevaluate our moral tenets and religious beliefs. It gives us access to more information about ourselves then we really want to know or, especially, to share with others. Humans demonstrate a need for some mystery in their lives and will invent the requisite amount when necessary. Knowing your genetic potential, although it reflects only a possibility, is unsettling, particularly if people start to act on the basis of that information. There seems for most people to be a comfort zone in which some things must be left to chance to give a person the opportunity to, or at least the feeling that they can, control their own destiny. The biggest and most far-reaching impact of genetic engineering will be social: How can we guarantee the privacy of the individual? Living in the age of genetic responsibility will not be easy. As is increasingly common in modern society, there is too much information, too many choices, and too many decisions to be made.

References

Adam, S., et al. "Five Year Study of Prenatal Testing for Huntington's Disease: Demand, Attitudes, and Psychological Assessment." *Journal of Medical Genetics* 30 (1993): 549–556.

Aldridge, S. "Researchers Tackling Bioremediation Challenge in Eastern Europe." *Genetic Engineering News* (October 1, 1997): 1.

Barwell, B. E., and R. J. Pollitt. "Attitudes des parents vis- à-vis du diagnostic prenatal de la phenylcetonurie." *Archives of French Pediatrics* 44 (1987): 665–666.

Biotechnology and the European Public Concerted Action Group (BEPCAG). "Europe Ambivalent on Biotechnology." *Nature* 387 (1997): 845–847.

Bureau of Justice Statistics. *A Survey of Recidivism among Prisoners Released in 1983.* Washington, DC: U.S. Department of Justice, March 1989.

D. Glass Associates, Inc. *Genetic Engineering News* 17 (1997): 43.

Dibner, M. D. *Genetic Engineering News* (1998, January 1): 34.

Eisiedal, E. F. *Biotechnology and the Canadian Public: Report on a 1997 Survey and Some International Comparisons.* Calgary, Canada: University of Calgary, 1997.

Elvers-Krebooms, G. "Predictive Testing for Huntington's Disease in Belgium." *Journal of Psychosomatic Obstetrics and Gynecology* 11 (1990): 61–72.

Flechas, F. W., and M. Latady. "Regulatory Evaluation and Acceptance Issues for Phytotechnology Projects." *Advances in Biochemical Engineering and Biotechnology* 78 (2003): 171–185.

"Genes, Social Science Survey." Public Opinion Online Database. Storrs, CT: Roper Center for Public Research, September 1990.

Hennen, L., T. Petermann, and J. Schmitt. "TA-Projekt 'Genomanalyse'— Chancer und Risiken Genetische Diagnostik" (TAB-Arbeitsbericht, No. 18, Bonn, 1990).

Hoban, T. J., and P. A. Kendall. *Consumer Attitudes about the Use of Biotechnology in Agriculture and Food Production.* Raleigh: North Carolina State University: 1992.

Lajos, I., and A. Czeizel. "Letter to the Editor: Reproductive Choices in Hemophiliac Men and Carriers." *American Journal of Medical Genetics* 28 (1987): 519–520.

Lee, T. R., C. Cody, and E. Plastow. *Consumer Attitudes towards Technological Innovations in Food Policy.* Guildford, England: University of Surrey, 1985.

Macer, D. *Attitudes to Biotechnology in New Zealand and Japan in 1997 (Eurobarometer Survey).* Christchurch, New Zealand: Eubios Ethics Institute, 1998.

March of Dimes. *Genetic Testing and Gene Therapy, National Survey Findings.* White Plains, NY: Louis Harris and Associates, 1992.

Moorish, D. L., and D. Abuelo. "Counseling Needs and Attitudes toward Prenatal Diagnosis and Abortion in Fragile-X Families." *Clinical Genetics* 33 (1988): 349–355.

National Science Board. *Science and Engineering Indicators—2002.*

Office of Technology Assessment. *New Developments in Biotechnology: Public Perceptions of Biotechnology* (OTA-BP-BA-45). Washington, DC: U.S. Government Printing Office, 1987.

PR Newswire. http://www.prenewswire.com, January 7, 1998.

Runge, C. F., and B. Ryan. "The Economic Status and Performance of Plant Biotechnology in 2003: Adoption, Research, and Development in the United States." Washington, DC: Council of Biotechnology Information, 2004.

Shamblott, M. J., J. Axelman, S. Wang, E. M. Bugg, J. W. Littlefield, P. J. Donovan, P. D. Blumenthal, G. R. Huggins, and J. D. Gearhart. "Derivation of Pluripotent Stem Cells from Cultured Human Primordial Germ Cells." *Proceedings of the National Academy of Sciences of the United States of America* 95 (1998): 13726–13731.

A Survey of the Use of Biotechnology in U.S. Industry. Washington, DC: U.S. Department of Commerce, October, 2003.

Thomson, J. A., J. Itskovitz-Eldor, S. S. Shapiro, M. A. Waknitz, J. J.

Swiergiel, V. S. Marshall, and J. M. Jones. "Embryonic Stem Cell Lines Derived from Human Blastocysts." *Science* 282 (1998): 1145–1147.

U.S. EPA. "Introduction to Phytoremediation." EPA document 600/R-99/107. EPA National Risk Management Research Laboratory, 1999.

U.S. General Social Survey. Chicago: National Opinion Research Center, 1990.

Wagner, T. "License Granted for Human Cloning." Associated Press. *Lexington Herald-Leader,* February 9, 2005.

Wilmut I., et al. "Viable Offspring Derived from Fetal and Adult Mammalian Cells." *Nature* 385 (1997): 810–830.

Wong, J. F. "Is Biotech in the Midst of a Fifty-Year Cycle?" *G&E News* 25, 5 (2005): 60.

7

Directory of Organizations, Associations, and Agencies

This chapter describes selected U.S. and international organizations connected with medical genetics, genetic engineering, or the impact of the technology on society. Many more are available online. Brief background information on the organization; the purpose of the group; its address, Web site, and contact information; the services provided; and publications of the organization are included. For the most part, these are organizations that are open to the general public. Organizations for professionals engaged in various aspects of genetic engineering have been omitted. If you are looking for information on a genetic disease, the National Organization for Rare Diseases maintains a comprehensive listing and links at their Web site (http://www.rarediseases.org).

Alliance of Genetic Support Groups
4301 Connecticut Avenue NW, Suite 404
Washington, DC 20008
(202) 966-5557; (800) 336-GENE (4363)
FAX: (202) 966-8553
E-mail: info@geneticalliance.org
Web site: http://www.geneticalliance.org
Contact: Executive director

The alliance, founded in 1985, comprises volunteer genetic organizations, professionals, and interested individuals who support people with genetic disorders and their families by assisting interactions with government agencies, healthcare professionals, and service providers. It provides technical assistance and information for referrals to the support groups and the public. The group maintains a database for its membership and offers an annual recognition award.

PUBLICATIONS: *Alliance Alert* (monthly newsletter), *Alliance Health Insurance Resource Guide, Directory of Voluntary Genetic Organizations and Related Resources, Informed Consent: Participation in Genetic Research Studies, Media Reporting in a Genetic Age, Alliance Resource Guide on Peer Support Training Programs*

American Genetic Association
P.O. Box 257
Buckeystown, MD 21717-0257
(301) 695-9292
FAX: (301) 695-9292
E-mail: agajoh@mail.ncifcrf.gov
Web site: http://www.theaga.org

A largely professional organization founded in 1903 includes biologists, zoologists, geneticists, botanists, and others exploring basic and applied research in genetics. Formerly called American Breeders Association.

PUBLICATION: *Journal of Heredity*

American Parkinson Disease Association
135 Parkinson Avenue
Staten Island, NY 10305
(718) 981-8001; (800) 223-2732
FAX: (718) 981-4399
E-mail: apdas@apdaparkinson.org
Web site: http://www.adpaparkinson.org
Contact: Vice president

Since 1961, the association has worked to alleviate the suffering of affected individuals and their families by subsidizing information and referral centers and funding research into cures for the disease. Counseling services are offered to patients and families through 51 information and referral centers. It maintains a library, not open to the public, and offers various awards for contributions to patient well-being and contributions to research on Parkinson's disease.

PUBLICATIONS: *American Parkinson Disease Association Newsletter* (quarterly), booklets on dealing with Parkinson's disease

American Society of Human Genetics
9650 Rockville Pike
Bethesda, MD 20814-3998
(301) 634-7300; 866-HUM-GENE
E-mail: society@ashg.org
Web site: http://genetics/faseb.org/genetics/ashg
Contact: Executive director

The American Society of Human Genetics, founded in 1948, is one of the oldest professional genetics organizations. Its members are involved in various aspects of medical genetics from basic research to clinical applications
    PUBLICATIONS: *American Journal of Human Genetics, Guide to Human Genetics Training Programs in North America, Enhancement of K–12 Human Genetics Education*

American Society of Plant Physiologists
15501 Monona Drive
Rockville, MD 20855-2768
(301) 251-0560
FAX: (301) 279-2996
E-mail: info@aspb.org
Web site: http://www.aspb.org
Contact: Executive director

The American Society of Plant Physiologists is a professional organization of scientists interested in plant physiology. It is a source of plant expertise including traditional biology and genetic engineering.
    PUBLICATION: *Plant Physiology, Agricultural Ethics in a Changing World, Plants, Genes, and Crop Biotechnology* (2nd edition)

Applied Research Ethics National Association (ARENA)
126 Brookline Avenue, Suite 202
Boston, MA 2115
(617) 423-4112
FAX: (617) 423-1185
E-mail: info@arena.org
Web site: http://www.primr.org
Contact: President

ARENA was founded in 1986 and is now part of Public Responsibility in Medicine and Research (PRIM&R). It consists of researchers, administrators, and professionals interested in bioethics, including members of institutional review boards, hospital ethics committees, patient advocacy groups, and institutional animal care and use groups. The group monitors federal legislation and public policy issues and provides expert consultation to members. They maintain a reference library and computerized mailing lists.

PUBLICATIONS: *ARENA Newsletter* (quarterly) and various brochures, educational programs to deal with HIPAA (Health Insurance Portability and Accountability Act) regulations.

A-T Medical Research Foundation
5241 Round Meadow Road
Hidden Hills, CA 91302
(818) 704)-8146
FAX: (818) 704-8310
E-mail: Becca4435@aol.com
Web site: http://www.gspartners.com/at
Contact: President

This multinational organization established in 1983 funds research on ataxia-telangiectasia, a genetic neurodegenerative disease. The group maintains a reference library of clippings and articles. It also awards a grant to support research.

PUBLICATION: *A-TMRF Newsletter*

Bioindustry Association
14/15 Belgrave Square
London SW1X 8PS England
+44(0) 20 7565 7190
FAX: +44(0) 20 7565 7191
E-mail: admin@bioindustry.org
Web site: http://www.bioindustry.org
Contact: Executive director

A national organization of individuals interested in biotechnology development in the United Kingdom, established in 1989. It seeks out contacts with trade associations and biotechnology organizations worldwide and represents member interests in regulatory affairs, government policy, and financing. In support of their mission, the association conducts educational workshops, seminars, and conferences and maintains a reference library, a computer database, and e-mail facilities.

PUBLICATIONS: *Bioscience UK* (annual), *BIA Annual Review,* electronic newsletter and reports (weekly)

Biotechnology Industry Organization
1225 Eye Street NW, Suite 400
Washington, DC 20005
(202) 962-9200
E-mail: info@bio.org
Web site: http://www.bio.org
Contact: President

BIO, founded in 1993, is an organization of industrial firms in the United States engaged in human healthcare, animal science, and chemical and industrial fermentation production. It also includes equipment manufacturers and servicers involved with recombinant DNA or hybridoma/immunologic technologies. Its mission is to provide information on biotechnology issues regarding various problems facing the industry such as U.S. and international regulations, patents, and financing opportunities. Much of this information is accessible through its Web site. The group interacts with Congress, a variety of government and nongovernment agencies, regulatory bodies, and the public. BIO maintains a resource library and provides a recognition award.
PUBLICATIONS: *BIO Bulletin, BIO News* online.

Birth Defect Research for Children
930 Woodcock Road, Suite 225
Orlando, FL 32803
(407) 895-0802
Web site: http://www.birthdefects.org
Contact: Executive director

A multinational organization (formerly Association of Birth Defect Children) created in 1982 brings together parents, healthcare professionals, other individuals, and organizations and services concerned with birth defects, particularly those associated with environmental exposure. BDRC conducts an educational campaign warning of the effects of prenatal exposure to environmental agents; sponsors, studies, and compiles statistics; and monitors and engages in legislative activities. The association maintains a computer database, a National Birth Defect Registry, and a parent matching service.

PUBLICATIONS: *Birth Defect News* (monthly e-newsletter), *Why My Child?*, fact sheets, and videotapes

Chromosome 18 Registry and Research Society
7155 Oakridge Drive
San Antonio, TX 78229
(210) 657-4968
FAX: (210) 657-4968
Web site: http://www.chromosome18.org
Contact: President

This multinational organization, founded in 1990, strives to locate individuals with chromosome 18 disorders and to educate the families and the public about prognosis and treatment. The group seeks to encourage research in the area and to link affected families and their physicians to the research community. It funds clinical research and maintains a parent network.

PUBLICATION: *Chromosome 18 Communique* (quarterly newsletter)

Council for Responsible Genetics
5 Upland Road, Suite 3
Cambridge, MA 02140
(617) 868-0870
FAX: (617) 491-5344
E-mail: crg@gene-watch.org
Web site: http://www.gene-watch.org
Contact: Executive director

This multinational watchdog group founded in 1983 monitors the biotechnology industry with concern for the social implications of genetic technology development. Primary issues of concern are genetically engineered foods, military uses of biotechnology, cloning and human genetic manipulation, and issues of genetic discrimination. The council maintains a reference library and a speakers bureau.

PUBLICATION: *GeneWatch* (bimonthly newsletter)

Crop Science Society of America
677 S. Segoe Road
Madison, WI 53711
(608) 273-8080
FAX: (608) 273-2021

E-mail: info@a-s-f.org
Web site: http://www.crops.org
Contact: Executive vice president

This is a professional society of individuals interested in issues surrounding plants and their cultivation as crops.
PUBLICATIONS: professional journals, e-newsletter

European Federation of Biotechnology
Pg. Lluis Companys 23
08010 Barcelona Spain
+34 93 268 7703
FAX: +34 93 268 4800
E-mail: info@efb-central.org
Web site: http://www.efb-central.org
Contact: General Secretary

The EFB is a multinational organization representing technical and scientific organizations in 26 European and non-European countries. Its purpose is to further biotechnology. Founded in 1978, it is concerned with animal and plant cell culture, production and processing of biomaterials, and environmental and safety issues.
PUBLICATIONS: *Annual Report of the EFB, EFB Newsletter* (English), *Proceedings of European Congress of Biotechnology* (periodic), fact sheets, briefing papers, review papers

Fanconi's Anemia Research Fund, Inc.
1801 Williamette
Eugene, OR 97401
(541) 687-4658; (800) 828-4891
FAX: (541) 687-0548
Web site: http://www.fanconi.org
Contact: Family support contact

Since 1989, the research fund has provided support and networking advice to families, and updates them on new research findings. It operates a support group and conducts fundraisers for research. The organization maintains a library open to the public.
PUBLICATIONS: *FA Family Directory, FA Family Newsletter* (semiannual), *Fanconi Anemia: A Handbook for Families and Their Physicians*

Foundation on Economic Trends
4520 East West Highway, Suite 600
Bethesda, MD 20814
(301) 656-6272
FAX: (301) 654-0208
E-mail: dhjohnston@foet.org
Web site: http://www.foet.org
Contact: Director

The foundation serves as a watchdog on technology and society issues in general and has entered on occasion into litigation to challenge government and industry policies. Founded in 1977 by Jeremy Rifkin, a noted activist, it has supported a wide variety of his causes.

Henry A. Wallace Institute for Alternative Agriculture
Winrock International
1621 N. Kent Street, Suite 1200
Arlington, VA 22209
(703) 525-9430
Web site: http://www.winrock.org
Contact: Executive director

Founded in 1983, this organization bands together farmers, agriculture scientists, educators, nutritionists, and consumers seeking to promote production methods that are economically viable, environmentally sound, and sustainable through development of educational programs and scientific research. It monitors and reports on related government activities.

PUBLICATIONS: *Alternative Agriculture News* (e-newsletter), *American Journal of Alternative Agriculture* (quarterly), reports, policy studies

International Centre for Genetic Engineering and Biotechnology
AREA Science Park Padriciano 99
I-34012 Trieste Italy
+(39) 040 37571
FAX: +(39) 040 226555
E-mail: icgeb@icgeb.org
Web site: http://www.icgeb.trieste.it
Contact: Director

This multinational organization, founded in 1984, promotes the use of biotechnology to solve problems of developing nations. It sponsors training programs as well as research and development in healthcare, agriculture, and industrial applications of biotechnology. The centre advises on biotechnology issues and maintains a reference library, a database, and electronic mail.

PUBLICATIONS: *Activity Report* (annual), *Helix* (quarterly newsletter in English)

March of Dimes Birth Defects Foundation
1275 Mamaroneck Avenue
White Plains, NY 10605
(914) 428-7100
Fax: (914) 428-8203
Web site: http://www.marchofdimes.com
Contact: President

Founded in 1938 by President Franklin D. Roosevelt, the March of Dimes promotes prevention of birth defects through proper prenatal care by addressing mother-child health issues such as low birth weight and maternal substance abuse.

PUBLICATION: *Genetics in Practice* (monthly e-newsletter); a separate Web site (http://www.modimes.org) provides educational material on genetic diseases, including spina bifida, sickle cell disease, the neurofibromatoses, cleft lip and palate, clubfoot, and Down syndrome

Michael Fund (International Foundation for Genetic Research)
4371 Northern Pike
Pittsburgh, PA 15146
(412) 374-0111
E-mail: chuckdet@michaelfund.org
Web site: http://www.michaelfund.org
Contact: Executive director

The fund, with its pro-life stance, has supported research on Down syndrome and related genetic disorders since 1978. The group opposes prebirth detection of disorders as well as abortion or euthanasia of afflicted children or adults. As an alternative, it seeks to improve care, treatment, and rehabilitation of afflicted children and adults through the support of professionals and public advocates.

PUBLICATION: *TMF Newsletter*

National Foundation for Jewish Genetic Diseases
250 Park Avenue, Suite 1000
New York, NY 10177
Center for Jewish Genetic Diseases, Mt. Sinai School of
  Medicine
Web site: http://www.nfjgd.org
(212) 371-1030
Contact: President

The foundation, established in 1974, seeks to have an impact on the seven known genetic diseases affecting children of mainly Ashkenazi Jewish heritage (Gaucher's disease, dysautonomia, Tay-Sachs disease, Bloom's syndrome, Niemann-Pick disease, and mucolipidosis IV). It supports basic medical research and acts as a referral agency for individuals. The organization promotes the establishment of disease carrier identification, provides genetic counseling including prenatal detection, and conducts a national education campaign.

PUBLICATIONS: a variety of books and brochures on Fanconi anemia, Tay-Sachs disease, β-lipoproteinemia, Bloom's syndrome, Canavan disease, familial dysautonomia, Gaucher's disease, mucolipidosis IV, Niemann-Pick disease, and Torsion dystonia

National Foundation for Rare Diseases
55 Kenosia Avenue
Danbury, CT 06813-1918
(203) 744-0100; (800) 999-6673 (voicemail)
TDD: (203) 797-9590
FAX: (203) 798-2291
E-mail: orphan@rarediseases.org
Web site: http://www.rarediseases.org

The foundation was established in 1983 by a group of patients and families who worked together to get the Orphan Drug Act through Congress. The act provides incentives for development of therapeutics for diseases afflicting fewer than 200,000 individuals. There are more than 6,000 human genetic diseases in this category, and in some instances only a few cases of the diseases occur each year worldwide. The mission of this nonprofit nongovernmental organization is to help people, assist organizations, and influence public policy through education, advocacy, and research. The foundation maintains a database of more than 1,150 diseases and an index of more than 2,000 organizations concerned with individual diseases with many links at their Web site.

National Fragile X Foundation
P.O. Box 190488
San Francisco, CA 94119
(925) 938-9300; (800) 688-8765
FAX: (925) 938-9315
E-mail: NATLFX@FragileX.org
Web site: http://www.fragilex.org
Contact: Executive director

Founded in 1974, the foundation is a multinational organization uniting professionals and families to support research and improve the treatment of this X chromosome–linked genetic disease causing mental retardation in affected males. It supports, advises, and assists parents of children with fragile X syndrome. The organization maintains a reference library.

PUBLICATIONS: *National Fragile X Foundation Newsletter* (quarterly), audio- and videotapes, brochures, and information packets for families and professionals

National Health Federation
P.O. Box 688
Monrovia, CA 91017
(626) 357-2181
FAX: (626) 303-0642
E-mail: contact-us@thenhf.com
Web site: http://www.thenhf.com
Contact: President

This large, established, multinational organization, founded in 1955, seeks alternatives to "organized medicine, the pharmaceutical industry, and other special interests" in support of individual freedom of choice in matters of health. The group serves as watchdog and attempts "corrective" measures through education, legislation, and coordination of like-minded organizational efforts. The federation supports research in such areas as laetrile testing and conducts legislative lobbying. It maintains a computer database and a public library of some 25,000 entries on alternative health and related laws.

PUBLICATION: *Health Freedom News* (quarterly)

National Institute for Science, Law, and Public Policy
1400 16th St. NW, Suite 101
Washington, DC 20036
(202) 462-8800
FAX: (202) 265-6564
Web site: http://swankin-turner.com/nislapp.html
Contact: President

The institute was founded in 1978 to influence public policies on sustainable agriculture, food safety, and nutrition. It promotes technologies avoiding chemical fertilizers and pesticides. The organization operates an information clearinghouse on aspartame, federal regulatory practices on milk pricing, use of prescription drugs, and interpretation of food and drug law. Training is provided through internships in areas of concern. The group provides information on request and maintains a speaker's bureau.

PUBLICATIONS: *The Earth and You Eating for Two, Eat Wise, Healthy Harvest IV: A Directory of Sustainable Agriculture Organizations* (annual)

National Neimann-Pick Disease Foundation
P.O. Box 49
401 Madison Avenue, Suite B
Ft. Atkinson, WI 53538
(920) 563-0930; (877) 287-3672
FAX: (920) 563-0931
E-mail: nnpdf@idcnet.com
Web site: http://www.nnpdf.org
Contact: Chairman

This multinational organization of parents, friends, and health and education professionals to promote research on Niemann-Pick disease was founded in 1992. This genetic disease affects sphingomyelin lipid and cholesterol metabolism. The foundation provides medical and educational information, offers support to families of children with the disease, supports legislation, and sponsors research on Niemann-Pick disease. It also maintains a reference library and a phone referral service.

PUBLICATION: *Niemann-Pick Newsletter* (quarterly)

National Society of Genetic Counselors
233 Canterbury Drive
Wallingford, PA 19086-6617

(610) 872-7608 (voice mail)
FAX: (610) 872-1192
E-mail: FY@nsgc.org
Web site: http://www.nsgc.org
Contact: Executive director

Founded in 1979, this multinational professional organization comprises genetic counselors and people with an interest in genetic counseling. A major focus is on education of the genetic counselor and promotion of the field in view of the increased need for such services with the advent of new genetic tests. The group provides a speaker's bureau and compiles statistics related to genetic counseling. It maintains a computer job-matching service and a general database.

PUBLICATIONS: *Journal of Genetic Counseling* (quarterly), *Perspectives in Genetic Counseling* (quarterly), a variety of high school and college level career packets, and genetic discrimination resources.

Neurofibromatosis, Inc.
P.O. Box 18246
Minneapolis, MN 55418
(800) 942-6825
Web site: http://www.nfinc.org
Contact: President

A multinational support group founded in 1988 to serve individuals with neurofibromatosis, their parents, caregivers, and healthcare providers. Neurofibromatosis is a genetic disorder linked to a number of neurological disabilities in which tumors form on nerves. The organization informs local, state, and national legislators about the needs of neurofibromatosis families and supports and funds medical and sociological research, including an annual scholarship, on possible cures and treatment. Neurofibromatosis Inc. identifies peer-counseling and local support resources and offers referrals to medical resources. It maintains an archival library that is not open to the public.

PUBLICATIONS: *NF Ink* (newsletter, semiannual), *Understanding Neurofibromatosis: An Introduction for Patients and Parents* (booklet), *NF Handbook*

Physicians Committee for Responsible Medicine
5100 Wisconsin Avenue, NW, Suite 400

Washington, DC 20016-4131
(202) 686-2210
FAX: (202) 686-2216
E-mail: pcrm@pcrm.org
Web site: http://www.pcrm.org
Contact: President

The committee is a large lay organization of physicians, scientists, and healthcare professionals and interested others seeking to increase awareness of the importance of preventative medicine and nutrition. Founded in 1985, the group raises scientific and ethical questions about the use of animals and humans in medical research. Recent activity includes organizing a coalition of environmental and public health groups to push for adoption of a requirement for short-term nonanimal tests in lieu of those regulations mandating animal tests. The Physicians Committee maintains a speaker's bureau and both general and breast cancer hotlines.

PUBLICATIONS: *Alternatives in Medical Education, Food for Life, Good Medicine* (quarterly), *The Power of Your Plate,* brochures and fact sheets, video on Web site

Prader-Willi Syndrome Association U.S.A.
5700 Midnight Pass Road
Sarasota, FL 34242
(941) 312-0400; (800) 926-4797
FAX: (941) 312-0142
E-mail: national@pwsausa.org
Web site: http://www.pwsausa.org
Contact: Executive director

The association is a multinational organization of families and professionals, founded in 1975, that works to promote communication about this genetic syndrome, mainly how to cope with it. It supports research and establishment of treatment facilities, conducts educational programs, and compiles statistics. The group maintains a reference library and a computer database.

PUBLICATIONS: *Gathered View* (bimonthly newsletter); *Management of Prader-Willi Syndrome* (book); *Food, Behavior and Beyond* (DVD); *Prader-Willi Syndrome: Coping with the Disease* (book); *Prader-Willi Syndrome Is What I Have, Not Who I Am* (book); and various information packets

Public Responsibility in Medicine and Research
120 Brookline Avenue, Suite 202
Boston, MA 2115-3920
(617) 423-4112
FAX: (617) 423-1185
E-mail: info@primr.org
Web site: http://www.primr.org
Contact: Executive director

This is a multinational group of researchers, clinicians, administrators, attorneys, and laypersons supporting responsible animal and human research. The organization was founded in 1974 and is allied with the Applied Research Ethics National Association. It primarily educates the healthcare community about development of research regulation and constructively attempts to counter the public's increasingly hostile attitude toward scientific research. The group sponsors public forums for discussion of issues and acts as a resource for members in preparing presentations for legislative or other types of public hearings.

   PUBLICATIONS: *Conference Report* (semiannual proceedings), *Guidebook on Institutional Animal Care and Use Committees, Human Subjects Guidebook,* and a variety of educational materials

Rural Advancement Fund-USA (RAFI-USA)
P.O. Box 640
Pittsboro, NC 27312
(919) 542-1396
FAX: (919) 542-0069
Web site: http://www.rafiusa.org
Contact: Executive director

This multinational organization, founded in 1990, promotes conservation and the sustainable use of agriculture, family farms, and socially responsible use of new technologies.

   PUBLICATIONS: *The Community Seed Bank Kit* (preserving traditional crop varieties), *The Laws of Life: Another Development and the New Biotechnologies* (social and economic impact of new biotechnologies on third world), *RAFI Communique* (issues on biodiversity, biotechnology, and intellectual property), *Shattering: Food, Politics, and the Loss of Genetic Diversity* (governments and corporations struggle for control of access to world's plant genetic resources), *Farmer's Guide to GMO's,* monthly e-bulletins

Turner Syndrome Society of the United States
14450 T.C. Jester, Suite 260
Houston, TX 77014
(832) 249-9988; (800) 365-9944
FAX: (832) 249-9987
E-mail: tssus@turnersyndrome-us.org
Web site: http://www.turnersyndrome.org
Contact: Executive officer

Founded in 1987, the society seeks to link individuals suffering from the disease, their families, and healthcare professionals. Turner syndrome is a genetic disease affecting females, resulting in short stature and kidney, cardiac, and motor perception difficulties. The group promotes public awareness of the medical and sociopsychological impact of the syndrome on those with the disease and provides healthcare professional referral. It maintains a speaker's bureau, a reference library, and a computer database.

PUBLICATIONS: *Newsletter* (quarterly), *Turner's Syndrome: A Guide for Families*, *Turner's Syndrome: A Guide for Physicians*

World Aquaculture Society (USA)
Louisiana State University
143 J. M. Parker Coliseum
Baton Rouge, LA 70803
(225) 388-3137
FAX: (225) 388-3493
E-mail: wasmas@aol.com
Web site: http://ag.arizona.edu/azaqua/WAS/uschap.htm
          http://www.was.org (international)
Contact: Manager

The society, founded in 1970, is a multinational organization dedicated to the evaluation, promotion and distribution of scientific and technological advancement in marine sciences worldwide. It promotes education and technical training and disseminates information on issues in aquaculture and mariculture.

PUBLICATIONS: *Advances in World Aquaculture* (periodic), *Journal of the WAS* (quarterly), *World Aquaculture* (quarterly—science, technology, political and economic updates)

8

Selected Print and Nonprint Resources

his chapter provides a selected annotated bibliography of printed resources (books, reports, periodicals, and directories) on various aspects of genetic engineering. The books are grouped according to their main subject content—General (includes history), Ethics, International, Legal, Business, Agriculture, Environment, and Science—and then arranged alphabetically. Sources that disagree with the mainstream opinion are indicated with an asterisk. A group of books on genetic engineering for young adults and younger readers is also provided. Following this are a selection of U.S. government, congressional, agency, and nongovernment reports on applications of genetic engineering and its societal impact. A list of periodicals, newsletters, and directories provides an introduction to more current and often more focused information on scientific and social issues.

Books

Note: * indicates sources with an alternative opinion to the mainstream.

General

Aldridge, S. *The Thread of Life: The Story of Genes and Genetic Engineering.* New York: Cambridge University Press, 1996. 258p. ISBN 0-521-46542-7.

The focus of this book is primarily on the explanation of the cellular systems that molecular biologists have learned to control. By comparing human capabilities of recombinant DNA manipulation to nature's evolutionary and contemporary processes, the author provides useful perspective on their relative magnitudes. A section on how cellular DNA is organized in the genome and how that genetic information is regulated (as it is currently understood) gives an inkling of the complex program that the Human Genome Project hopes to elucidate. The relationship of genes and cancer provides a description of genetic predisposition from which lessons can be extrapolated to other diseases. An illuminating discussion of what a person's genetic identity means reveals that it differs from the public conception. Important for the educated lay reader, this particular section exposes the myth, often inferred in the media, that because certain simple things have been done, complete human or other organism engineering will be a trivial operation. Not only society but nature conspires against genetic engineers. Examples of uses to which biotechnology has been put in biopharmaceuticals, medicines, agriculture, industrial production as well as in environmental remediation and energy production are also discussed in some detail. The author leaves to the many other books a detailed treatment of ethical issues and the social impact of genetic engineering.

Alexander, B. *Rapture: How Biotech Became the New Religion.* Cambridge, MA: Perseus/Basic Books, 2003. 289p. ISBN 0-738-20761-6.

A consideration of biotechnology as a futurist culture and what the opposition of so-called bio-Luddites to the applications of stem cells, human cloning, and genetic engineering means about society.

Bains, W. *Biotechnology from A to Z.* New York: Oxford University Press, 1993. 358p. ISBN 0-199-63334-7.

A detailed introduction to biotechnology and its applications is provided at a level requiring more technical experience for a full understanding than *The Thread of Life* by Susan Aldridge (cited earlier). The potential consequences of application of the technology are also discussed in some detail. Genetic engineering, particularly as reflected in human cloning and gene therapy, and the numerous moral, ethical, and social issues that accompany these technologies receive less coverage, partly because these considerations became prominent after 1993 when the book was published.

Cherfas, J. *Man-Made Life: An Overview of the Science, Technology, and Commerce of Genetic Engineering.* New York: Pantheon Books, 1982. 270p. ISBN 0-394-52926-X.

This book recounts the history of the science of genetic engineering. It is conversationally written and identifies many of the scientists involved. The author provides lucid explanations of what the scientists were doing and why they were doing it during the development of the technology rather than dryly presenting the facts.

Department of Government Relations and Science Policy. *Biotechnology* (information pamphlet). Washington, DC: American Chemical Society. 1995. 16p.

A basic primer on biotechnology and its applications compressed into a small space.

DeSalle, R., and D. Lindley *The Science of Jurassic Park and the Lost World or, How to Build a Dinosaur.* New York: Basic Books, 1997. 194p. ISBN 0-465-07379-4.

Jurassic Park and *The Lost World,* those fantastic, thrilling Steven Spielberg–directed movies with dinosaurs re-created from preserved DNA by recombinant DNA technology, were a marvel of special effects and cinematic art. The story line of Michael Crichton's books on which these movies are based is similarly a marvel of science and science fiction, highly entertaining and seemingly believable. In the tradition of the great science fiction author Jules Verne, there is just enough fact to give the illusion of reality. Before getting too carried away about the possibilities of cloning ancient life, it's worth considering what the chances are that such events could really happen. *The Science of Jurassic Park* is a highly readable assessment of what series of highly unlikely events would have to occur, what would be required to bring them about, and under what set of conditions. Along the way the reader finds out quite a lot about recombinant DNA technology, its strengths and its weaknesses, the reality of cloning technology, the likelihood of dinosaur societies, and survival of disease.

Finally, the author considers the ethical questions of resurrecting an extinct creature. Even if it could be done, should it?

Fox, M. W. *Superpigs and Wondercorn: The Brave New World of Biotechnology and Where It All May Lead.* New York: Lyons and Burford, 1992. 209p. ISBN 1-558-21182-9.

A critical assessment of uncontrolled use of biotechnology and genetic engineering. Although not opposed in principle to the use of the technology, the author warns of the overselling of the possibilities. The text is liberally sprinkled with facts and figures supporting the theme of the book.

Fumento, M. *Bioevolution: How Biotechnology Is Changing Our World.* San Francisco: Encounter Books, 2003. 510p. ISBN 1-893-55475-9.

This book is an essay on the impact of the biotechnology revolution at multiple levels in the world economy and culture. It goes beyond the headlines, the moralist rhetoric, and the biotech hype to analyze how the advances have permeated the world culture. Once started, it was not a reversible process. The author is a journalist and a keen observer of biotechnology who has written weekly medical columns and for investment magazines. The book includes an extensive (more than 100 pages) series of notes and references.

Hellman, H. *Great Feuds in Technology: Ten of the Liveliest Disputes Ever.* New York: Wiley and Sons, 2004. 248p. ISBN 0-471-20867-1.

A member of the "Great Feuds in . . ." series of books. Although most of the book is devoted to other technologies, two chapters take up biotechnology confrontations: Chapter 9, "Venter vs. Collins—Decoding the Genome," and Chapter 10, "Rifkin vs. the Monsanto Company. Battling the Biotech World." The author provides a lively and highly readable discussion of the events surrounding the race to sequence the genome (Ch. 9) and social activism vs. business interests (Ch. 10).

Hodgson, J. G. *Bio-Technology: Changing the Way Nature Works.* London: Cassell, 1989. 124p. ISBN 0-304-31783-7.

Although somewhat outdated and lacking coverage of the newest trends, this book provides a first-rate introduction to biotechnology and its applications. Lavishly illustrated and filled with color photographs and computer models of the technology and the people at work in the field make this book a good place to get a visual feel and an understanding for the science.

*Hubbard, R., and E. Wald. *Exploding the Gene Myth: How Genetic Information Is Produced and Manipulated by Scientists, Physicians, Employers, Insurance Companies, Educators, and Law Enforcers.* Boston: Beacon Press, 1993. 206p. ISBN 0-807-00418-9.

As the book title suggests, Dr. Ruth Hubbard has a lot to say about the influence the genetic engineering revolution has or could have on society. The populist sentiment is evident, decrying the abuse of genetic information and the seemingly conspiratorial machinations of medicine, science, government, and industry combining to oppress certain individuals and minority groups. The number of alleged improprieties and potential ulterior motives detailed suggest an incredible scandal. Although there is probably a certain amount of nefarious activity, as there is in any human endeavor, only part of the story is told here, and the reader is left to judge the extent of the truth. Reality likely lies somewhere in between.

The technical background of the controversy and genetic concepts are explained in a facile manner. Despite having to impart a great deal of information, Hubbard presents her point of view in an engaging style aimed at the lay public. The book's notes and information resources give access to points of view often at odds with the scientific and government mainstream.

Judson, H. F. *The Eighth Day of Creation: Makers of the Revolution in Biology.* Plainview, NY: Cold Spring Harbor Laboratory Press, 1996. 714p. ISBN 0-879-69477-7.

This is a history of molecular biology that concentrates less on the science and more on the political and social repercussions of the science. The earlier edition of this book has not been available in the United States for a number of years. The new version provides additional material on some of the principal figures in the history and updates events since 1978.

*Kimbrell, A. *The Human Body Shop: The Engineering and Marketing of Life.* San Francisco: Harper, 1993. 348p. ISBN 0-062-50524-6.

This book takes an antitechnology view of the developments in organ transplant technology and genetic engineering. The author deals with a number of controversial issues concerning the patenting of life, use of body parts, and changing notions of self. Many of the arguments are made on the basis of what is "natural."

Kolata, G. B. *Clone: The Road to Dolly and the Path Ahead.* New York: William Morrow, 1998. 276p. ISBN 0-688-15692-4.

The author is the *New York Times* reporter who broke the story of the sheep Dolly, the first mammal cloned from differentiated tissue cells, to the American public. In this book, she presents her case in the style of scientific journalism rather than as a philosopher or as a self-proclaimed moralist. The history of the science follows the step-by-step process of the development of the technology through the widely accepted technology of cloned cattle embryos of prized stock to the virtually ignored creation of the twin lambs Megan and Morag from cloned embryonic cells, up to the highly publicized cloning of Dolly from a differentiated adult udder cell.

Kolata quotes a variety of opinions and in the end raises more questions than she answers. In the meantime, she explains how and why society has come to this point in the cloning issue clearly and understandably to a wide nontechnical audience and exposes some of the future implications of the technology and our reaction to it for further contemplation.

Krimsky, S. *Genetic Alchemy: The Social History of the Recombinant DNA Controversy.* Cambridge, MA: MIT Press, 1982. 445p. ISBN 0-262-11083-0.

A history of the early years of the development of recombinant DNA technology.

*Lear, J. *Recombinant DNA: The Untold Story.* New York: Crown, 1978. 280p. ISBN 0-517-53165-8.

John Lear, a veteran commentator on the communication between science and society, presents an expose-style view of the history of the recombinant DNA controversy. He weaves an engrossing tale of personalities and motives in which he detects a conspiracy of scientists to deceive and control the public. This highly readable book, written on a Ford Foundation grant, presents an alternative critical picture of how science and social issues interact. It is particularly instructive to consider these thoughts after the more than 20 years of genetic engineering practice that have passed since the publication of this book. Just what have society and scientists learned about each other in that time?

Levine, J. S., and D. Suzuki *The Secret of Life: Redesigning the Living World.* Boston: WGBH Television, 1993. 280p. ISBN 0-963-68810-3.

The book form of the popular *Nova* educational television series on genetic engineering.

Lim, H. A. *Genetically Yours: Bioinforming, Biopharmacy, and Bioforming.* River Edge, NJ: World Scientific, 2002. 417p. ISBN 9-810-24938-1.

The author discusses the implications of the biotechnology and genetic revolutions for individuals in having and using knowledge of their genetic background and what impact that will have on their life choices and medical treatment options.

Lyon, J., and P. Gorner. *Altered Fates.* New York: W. W. Norton, 1995. 636p. ISBN 0-393-31528-2.

Written by two *Chicago Tribune* journalists who won the Pulitzer Prize for reporting in 1987, this book provides an enlightening and entertaining treatment of the history of gene therapy and the development of the Human Genome Project. It includes discussions of gene repair and work with embryos. After making the case for the impact of genetic diseases, the authors make it clear they believe that society is not ready to handle the implications of gene therapy.

McGee, G. *Beyond Genetics.* New York: William Morrow/ HarperCollins, 2003. 231p. ISBN 0-060-00800-8.

The author describes the genome sequencing efforts and what it means to have that knowledge. He suggests how individuals will be using this new information to make decisions about food, medical treatment, family planning, and other issues in their daily lives.

Paula, L., ed. *Biotechnology for Non-specialists — A Handbook of Information Sources.* EFB Task Group on Public Perceptions of Biotechnology, 1997. 286p. ISBN 9-076-11001-8.

This handbook is designed to assist people with a broad range of interests and backgrounds in finding information in the public debate on biotechnology in Europe. Written information sources are cataloged, and lists and descriptions of organizations are provided along with a collection of Internet sites and acronyms. The book is intended as an information source and organizer rather than to provide detailed explanations of the science, which the author feels are found elsewhere.

Pence, G. E. *Cloning after Dolly: Who's Still Afraid?* Lanham, MD: Rowman and Littlefield, 2004. 211p. ISBN 0-742-53408-1.

The author tries to assess how human and other animal cloning will affect us in the long run. Whereas most books of this type dwell on the negative potential, he considers that because the technology is available, one would do well to consider neutral and even positive aspects. He is particularly interested in how it will change our way of thinking about ourselves once we get beyond knee-jerk reflexes against change. He corrects a number of common misconceptions, arguing, for example, that cloning will not affect genetic diversity in any substantial way.

Piller, C., and Yamamoto, K. R. *Gene Wars: Military Control over the New Genetic Technologies.* New York: Beech Tree Books, 1988. 302p. ISBN 0-688-07050-7.

A somewhat dated account of the involvement of the military with genetic engineering and recombinant DNA technologies.

*Rifkin, J. *The Biotech Century: Harnessing the Gene and Re-making the World.* New York: Jeremy P. Tarch/Putnam, 1998. 271p. ISBN 0-874-77909-X.

As he suggests by the title, Jeremy Rifkin has a few comments to make about the genetic engineering revolution that he opposed more than 20 years ago. The specter of human cloning (by nuclear transfer) has reawakened the original concerns of the early antigenetic engineering activists. Along the way, Rifkin reaffirms his Luddite leanings about the uses and abuses of technology in his examples of the acquisition of fire, the early printing press, the English law on common land, and finally passing on to what he perceives as our failing modern industrial workplace. His message to consider the ethical quandaries of technology in general, and those of biotechnology and genetic engineering in particular, comes through as it did in *Algeny*, although with less force and alarm than in the past.

*Rifkin, J., and N. Perlas. *Algeny.* New York: Viking Press, 1983. 298p. ISBN 0-670-10885-5.

In this book, Rifkin presents a critique of the then-emergent biotechnology epoch. The term *algeny*, newly coined by the authors, is drawn in parallel to alchemy, which in the ancient world was the culmination of the natural conversion of lesser materials into gold or perfection through the agency of fire. Biotechnology is cast as an analogous philosophy in which technology provides the molding force for change to the detriment of natural systems, spiced throughout by Rifkin with a hint of conspiracy.

This worldview in which technology is destined to destroy the natural order of the relationship of humankind to the earth is the driving force of Rifkin's actions in opposition to biotechnology in line with his many publications on ecological, social, and political issues.

Russo, E., and D. Cove *Genetic Engineering: Dreams and Nightmares.* New York: W. H. Freeman/Spektrum, 1995. 243p. ISBN 0-716-74546-1.

A detailed lay presentation of the promises and disasters of genetic engineering.

*Shiva, V., and I. Moser *Biopolitics: A Feminist and Ecological Reader on Biotechnology.* Atlantic Highlands, NJ: Zed Books. 1995. 294p. ISBN 1-856-49335-0.

A nontraditional view of biotechnology.

Shreeve, J. *The Genome War: How Craig Venter Tried to Capture the Code of Life and Save the World.* New York: Knopf, 2004. 403p. ISBN 0-375-40629-8.

A highly readable history of the race to sequence the genome and the high-stakes personalities involved.

Thompson, L. *Correcting the Code: Inventing the Genetic Cure for the Human Body.* New York: Simon and Schuster, 1994. 378p. ISBN 0-671-77082-9.

Larry Thompson, a master's level molecular biologist and a former science writer for the *Washington Post,* tells the history of genetic therapy, from the initial grandiose plans to the realization that developing the process would be long and hard and finally to the first true genetic treatment of four-year-old Ashanti De-Silva in September 1990. He vividly portrays the very personal stake the genetic pioneers had in their science, the disappointments, and the temptations, set in the background of the early days of the genetic revolution.

Watson, J. D. *The Double Helix.* New York: W. W. Norton, 1980. 298p. ISBN 0-393-01245-X.

The controversial story of the discovery of the helical structure of DNA as seen through the eyes of James Watson, Nobel laureate. Watson shatters the image of the white-coated calculating scientist toiling alone in a laboratory with his account of the coincidences, emotions, and conflict of strong personalities with different motives. Accounts by others involved in the search for the Holy Grail of genetics show the personal nature of science from diverse perspectives. A classic.

Watson, J. D., and A. Berry. *DNA: The Secret of Life.* New York: Knopf, 2003. 446p. ISBN 0-375-41546-7.

This coffee table–style book with many high-quality pictures is a historical narrative of the history of DNA by one of the pioneers.

Watson, J. D., and J. Tooze. *The DNA Story: A Documentary History of Gene Cloning.* San Francisco: W. H. Freeman, 1981. 605p. ISBN 0-716-71292-X.

This engaging history of the early days of the recombinant DNA story provides insight into the scientific, political, and legislative developments during that period. Reproduction of many letters exchanged between key scientists, administrators, and legislators brings this era to life by giving a view of the process rather than merely the final result. Reproductions of key portions of government documents and articles from newspapers, magazines, and science journals provide the societal background throughout.

Yount, L. *Biotechnology and Genetic Engineering.* New York: Facts on File, 2000. 280p. ISBN 0-816-04000-1.

A handbook of facts and issues of biotechnology.

Ethics

Andrews, L. B. *Future Perfect: Confronting Decisions about Genetics.* New York: Columbia University Press, 2001. 264p. ISBN 0-231-12162-8.

A discussion of the issues confronting how decisions about the availability and use of genetic information are being made.

Andrews, L. B., J. E. Fullarton, N. A. Holtman, and A. G. Motulsky, eds. *Assessing Genetic Risks: Implications for Health and Social Policy.* Washington, DC: National Academy Press, 1994. 338p. ISBN 0-309-04798-6.

This is a report by the Institute of Medicine gleaned from a series of workshops and meetings, public and private, on genetic testing and its impact on patients, providers, and the laboratories providing the services. Each section finishes with a series of conclusions and recommendations of the Committee on Assessing Genetic Risks. A unique contribution of this book is its consideration of what is required on a practical level to achieve the accuracy and accountability of the laboratory science (Ch. 3), public education to allow informed decision making (Ch. 5), and training of genetic professionals (Ch. 6) who will be presenting the options. They also consider the usual social, ethical, and legal implications of genetics policies. This is an informative but densely written book.

Bodner, W., and R. McKie. *The Book of Man: The Human Genome Project and the Quest to Discover Our Genetic Heritage.* New York: Simon and Schuster, 1995. 272p. ISBN 0-195-11487-6.

A description of the Human Genome Project and its implications for society.

Cole-Turner, Ronald. *The New Genesis: Theology and the Genetic Revolution.* Louisville, KY: Westminster/John Knox Press, 1993. 128p. ISBN 0-664-25406-3.

The author discusses genetic engineering and its uses in the context of humankind and Christian theology. He notes that in general, except for certain groups who oppose any technological intervention, little resistance is shown to medical applications. More concern is shown for any attempt to "improve" humans, although he acknowledges that medical necessity and "improvement" can converge on occasion.

Drlica, K. A. *Double-Edged Sword. The Promises and Risks of the Genetic Revolution.* Reading, MA: Addison-Wesley Longman, 1994. 256p. ISBN 0-201-40838-4.

The author concentrates primarily on the social and ethical consequences of the use of the influx of genetic information. He gives a list of genetic services providers and volunteer services for different genetic diseases.

Frankel, M. S., and A. Teich, eds. *The Genetic Frontier: Ethics, Law, and Policy.* Washington, DC: American Association for the Advancement of Science, AAAS Publication #93-27S1994, 240p. ISBN 0-871-68526-4.

This collection of articles on the implications of the Genome Project and genetic testing is distilled from an invitational conference of the AAAS on social aspects of genetic engineering. Contributors comment on defining the family, privacy, linking genetics and behavior, assigning responsibility, use in criminal justice, and intellectual property and patent rights versus human dignity.

Holdrege, C. *Genetics and the Manipulation of Life: The Forgotten Factor of Context.* Hudson, NY: Lindisfarne Press, 1996. 190p. ISBN 0-940-26277-0.

A philosophical discussion of genes and organisms.

Holland, S., L. Labacqz, and L. Zoloth, eds. *The Human Embryonic Stem Cell Debate: Science. Ethics, and Public Policy.* Cambridge, MA: MIT Press, 2001. 257p. ISBN 0-262-08299-3.

A collection of articles on the issues firing the debate over stem cells.

*Howard, T., and J. Rifkin. *Who Should Play God? The Artificial Creation of Life and What It Means for the Future of the Human Race.* New York: Delacorte Press, 1977. 272p. ISBN 0-440-09552-2.

This book considers the issues surrounding the increased control of genetics made feasible by recombinant DNA technology. The history of eugenics is recounted along with a description of modern reproductive technology and the impact the new technologies will have on social mores. The authors discuss recombinant DNA technology in terms of its use to detect genetic defects and to have the potential for genetic fixes such as gene therapies. The authors pose the question of whether these technologies are really needed or wanted by society. Examples of science and corporate abuses of power and mismanagement in other situations are cited to suggest that these should not be the groups that society allows to make decisions on the use of the new genetic technologies. The authors at the time they wrote the book were codirectors of the Peoples Bicentennial Commission. J. Rifkin is a well-known figure often found opposing the application of genetic technologies and involved with environmental issues.

Kevles, D. J., and L. Hood, eds. *The Code of Codes: Scientific and Social Issues in the Human Genome Project.* Cambridge, MA: Harvard University Press, 1992. 384p. ISBN 0-674-13645-4.

This collection of expert commentaries covers the spectrum of technologies that were required for the Human Genome Project to succeed and recounts the project's history through 1992. It also addresses many of the social issues that need to be confronted in the use of that information. When genetic testing should be applied and what it means for the individuals involved are the main themes of the book.

Kitcher, P. *The Lives to Come: The Genetic Revolution and Human Possibilities.* New York: Simon and Schuster, 1996. 381p. ISBN 0-684-80055-1.

The book is primarily directed at ethical issues, although some scientific issues are considered.

Kristol, W., and E. Cohen, eds. *The Future Is Now: America Confronts the New Genetics.* Lanham, MD: Rowman and Littlefield, 2002. 357p. ISBN 0-742-52195-8.

This book is a collection of essays, articles, and speeches addressing cloning and stem cell research focusing on their social impact. The arguments are pro or con, impassioned and reasoned, philosophical and pragmatic. It is meant to be a sampling of the discussions about the place of new genetic technologies and how they should be regulated to keep them from spinning out of control.

Lee, T. F. *Gene Future: The Promise and Perils of the New Biology.* New York: Plenum Press, 1993. 339p. ISBN 0-306-44509-3.

This book describes the many sides of the issues of the science and ethics of genetic engineering and the Human Genome Project.

Nelkin, D., and L. Tancredi. *Dangerous Diagnostics: The Social Power of Biological Information.* New York: Basic Books, 1989. 207p. ISBN 0-465-01573-5.

This book considers the consequences of biological testing and the effects of labeling individuals in various social contexts such as the workplace, in courts of law, and in healthcare.

O'Neill, T., ed. *Biomedical Ethics: Opposing Viewpoints.* San Diego, CA: Greenhaven Press, 1994. 312p. ISBN 1-565-10062-X.

Ethical issues surrounding the medical applications of high technology include some brought about by the genetic revolution.

Reiss, M. J., and R. Straughan. *Improving Nature: The Science and Ethics of Genetic Engineering.* Cambridge: Cambridge University Press, 1996. 288p. ISBN 0-521-45441-7.

A biologist and a moral ethicist consider the potential of the genetic engineering revolution and explain its pitfalls. This book supplies much-needed discussion on moral and ethical questions about the various aspects and applications of biotechnology. Pro and con viewpoints to each question are presented (without a conclusion) and are accompanied by critical facts and figures without vilifying either viewpoint. This evenhanded presentation emphasizes that there really are two (at least) sides to every issue. In doing so, such a format also points out how far apart the two sides remain.

In emphasizing the importance of public education in informing people to have meaningful discussions about biotechnology issues, the authors distinguish between education and information. Because the authors—a biologist (MJR) and a moral philosopher (RS)—are English, much of the information is from the European community, providing a unique viewpoint compared with most U.S. sources.

Resnik, D. B. *Owning the Genome: A Moral Analysis of DNA Patenting.* Albany: State University of New York Press, 2004. 235p. ISBN 0-791-45931-4.

A discussion of the controversy and intricacies of what it means to patent natural and modified DNA sequences. Should it be allowed, and if so, under what conditions? Who should be allowed to benefit?

Rothstein, M. A., ed. *Genetic Secrets: Protecting Privacy and Confidentiality in the Genetic Era.* New Haven, CT: Yale University Press, 1997. 511p. ISBN 0-300-07251-1.

The chapters in this compendium provide a comprehensive assessment of the impact of the new genetic technology and its medical and nonmedical uses on personal privacy. The "gen-etiquette" controversy and how it is being played out in the societal arenas of health and law and order are discussed in separate chapters by a group of clinical, scientific, legal, and ethicist experts. Particularly valuable is the inclusion of the international impact of the new technology and the issues being debated in countries other than the United States, with the focus on Europe. A working policy framework for protecting genetic information and personal privacy is proposed.

Ruse, M., and C. A. Pynes, eds. *The Stem Cell Controversy: Debating the Issues.* Amherst, NY: Prometheus Books, 2003. 308p. ISBN 1591020301.

A collection of articles on the issues of the stem cell debate.

Smith II, George P. *Bioethics and the Law: Medical, Socio-Legal and Philosophical Directions for a Brave New World.* Lanham, MD: University Press of America, 1993. 332p. ISBN 0-819-19177-9.

Bioethics is the point at which genetic engineering and biotechnology contact society's norms and expectations. The law attempts to put some consistent form to these sometimes-vague guidelines through legislation that proscribes which actions may be taken and which are not allowed. This book provides a discussion of this process.

Suzuki, D., and P. Knudtson. *Genethics: The Clash between the New Genetics and Human Values.* Cambridge, MA: Harvard University Press, 1990. 372p. ISBN 0-674-34566-5.

This is a book written by the television science communicator David Suzuki for the nonscientist. Dr. Suzuki provides a very readable discussion that begins by describing genes and DNA and then turns to the technology. The bulk of the book is concerned with the ethical implications of genetic technology. Topics covered include the blaming of aggressive behavior on chromosomes (XYY males), genetic screening, somatic and germ cell therapy, biological weapons, environmental damage to DNA, the value of genetic diversity, and the implications of crossing genetic boundaries with transgenic organisms. The authors raise numerous questions about the use of genetic technologies.

International

Brock, M. V. *Biotechnology in Japan.* The Nissen Institute/Routledge Japanese Studies series. New York: Routledge Press, 1989, 156p. ISBN 0-415-03495-7.

The policy-making process used in Japan is different from that in the United States. In Japan, a coalition comprising government bureaucracy, industry, and politics is essential. Such close ties would engender a congressional investigation in the United States. Japan is particularly attracted to biotechnology because it allows the efficient use of normally scarce resources in that island nation to reduce dependence on outside parties and economic forces. The shift to knowledge-intensive industries is significantly more advanced than in the United States. Most of the book is concerned with strategies for developing a biotechnology industry in Japan.

Doyle, J. C., and G. J. Persley, eds. *Enabling the Safe Use of Biotechnology: Principles and Practice.* Washington, DC: World Bank, 1996. 84p. ISBN 0-821-33671-1.

This book describes a global environmental agenda. The stated goal of the volume is to provide "a practical guide for policy-makers and research managers who are responsible for making decisions on ensuring the safe use of modern biotechnology." This report does not consider the global differences of opinion over introducing genetically engineered organisms into agriculture. The tenth in a series of monographs, Environmentally Sustainable Development Studies.

Juma, C., J. Mugabe, and P. Kameri-Mbote, eds. *Coming to Life: Biotechnology in African Economic Recovery.* London: Zed Books, 1995. 192p. ISBN 9-9966-41087-2.

The countries of Africa are experiencing economic, political, and ecological decline, all of which contribute to a decreasing ability to compete on the open world market. Their political, educational, and industrial institutional structures are unable to cope with changes such as a decreasing market for raw materials that have constituted the bulk of the exports of these nations. The situation is assessed in Cameroon, Ethiopia, Kenya, Tanzania, and Uganda, and a series of recommendations for African policy makers are made to facilitate the integration of new technology and institutional change to improve the economic outlook for the region.

Peritore, N. P., and A. K. Galve-Peritore. *Biotechnology in Latin America: Politics, Impacts and Risks.* Wilmington, DE: SR Books, 1995. 229p. ISBN 0-842-02556-1.

This collection of articles considers the general problems associated with developing nation economies and scientific and technical infrastructures. It looks specifically at the situation in Mexico, Cuba, and Columbia. Particularly relevant to these countries are the ideas of property rights (patent protection) balanced against gene pool rights (genetic property rights to natural diversity). Economic limitations for most Latin American nations have stunted the development of the appropriate infrastructure for technological advancement. This has led to the idea of converting national debt owed to developed nations into credits for technological advancement in developing nations, which then could afford products from developed nations.

Russell, A. M. *The Biotechnology Revolution: An International Perspective.* New York: St. Martins Press, 1988. 266p. ISBN 0-745-00013-4.

The emphasis of this book comes from the author's training in international relations. It provides a perspective on the process of development of regulations in the United Kingdom and Europe in particular, with some comments on Japan, West Germany, and Canada and little on the United States. The history of national developments and controls is reviewed. There is a fear that Western companies will test their genetically engineered products in underdeveloped countries, much as they sold less safe nuclear reactor designs there, because the undeveloped nations had less sophisticated regulatory standards.

Sasson, A. *Biotechnologies in Developing Countries: Present and Future. Vol. 1: Regional and National Survey.* Paris: UNESCO, 1993. 764p. ISBN 9-331-02875-8.

Sasson has been the director of the Bureau of Studies, Programming and Evaluation of UNESCO since 1974. This book provides a wealth of economic data and reports on biotechnological research and development activities. Overviews of achievements, expected developments, constraints on progress and application, research and industrial application strategies, consumer interest and biosafety, intellectual property, and regulation issues are presented, concluding with comments on the economic impact (with a set of forecast tables) and on how biotechnology is being pursued in developing countries.

Smith, J. E. *Biotechnology.* 3rd edition. Cambridge: Cambridge University Press, 1996. 236p. ISBN 0-521-44467-5.

This book considers the various aspects of genetic engineering particularly with respect to industrial applications. It covers the biotech industry, agriculture, biomining, concerns with patents and ethics, and environmental release of genetically modified organisms. The information is primarily from England, providing a distinctive European view of the subject.

Legal

Bernaer, T. *Genes, Trade, and Regulation: The Seeds of Conflict in Food Biotechnology.* Princeton, NJ: Princeton University Press, 2003. 229p. ISBN 0-691-11348-3.

The impact of genetic engineering on agricultural biotechnology in particular is a complex issue with the potential to distort local and global economics. The current regulatory structure is struggling to balance business interests with safety and market forces.

Billings, P. R., ed. *DNA on Trial: Genetic Identification and Criminal Justice.* Plainview, NY: Cold Spring Harbor Laboratory Press, 1992. 154p. ISBN 0-879-69379-7.

This book covers the history and impact of DNA forensic evidence on civil liberties and public policy. It contains an analysis of decisions made in various courts by juries considering DNA evidence.

Cook, T., C. Doyle, and D. Jabbari. *Pharmaceuticals, Biotechnology, and the Law.* New York: Stockton Press, 1991. 834p. ISBN 1-561-59040-1.

The laws respecting pharmaceuticals and biotechnology differ, sometimes significantly, among nations. This book reviews the law and legislation governing pharmaceuticals and biotechnology, concentrating on Great Britain and the European Economic Community (EEC) countries. The appendices contain the texts of the EEC Council Directives and other related documents of the late 1980s through the early 1990s.

Environmental Law Centre. *Law in the New Age of Biotechnology.* Edmonton, Canada: Environmental Law Centre, 1992. 220p. ISBN 0-921-50341-5.

This book describes the legal basis and present legislation governing biotechnology and environmental law in Canada.

Inman, K., and Rudin, N. *An Introduction to Forensic DNA Analysis.* Boca Raton, FL: CRC Press, 1997. 256p. ISBN 0-849-38117-7.

A discussion of modern methods of analyzing DNA and use of that information in courts of law. A technical source.

Laurie, G. T. *Genetic Privacy: A Challenge to Medicolegal Norms.* New York: Cambridge University Press, 2002. 335p. ISBN 0-521-66027-0.

The author is a senior lecturer in law at the University of Edinburgh. His analysis of the philosophical and legal meanings of privacy range from attempts to construct a legal framework to the practical workings in different societies. He feels that the new genetic technologies make it even more imperative that privacy rights be better defined than they have in the past and that the laws be brought into compliance with the outcome of this discussion. So far, he feels that most legislation has dealt with regulating misuse of genetic information in insurance coverage or compensation and in employment but leaves the deeper philosophical issues unaddressed. Legislation is also not uniform across countries, even at a baseline level.

Muth, A. S., ed. *Forensic Medicine Sourcebook.* Detroit, MI: Omnigraphics, 1999. 574p. ISBN 0-780-80232-2.

This book is a collection of articles and essays reprinted from various sources on the topic of forensic investigation—what is involved, the kind of evidence collected, how it is collected, and what the impact is or has been. Of importance to genetic engineering concepts are Chapter 27, "Automated DNA Typing"; Chapter 37, "The Potential of DNA Testing"; and Chapter 49, "Exonerated by DNA," all on the specific impact of DNA analysis admitted into evidence in criminal trials.

Shiva, V. *Biopiracy: The Plunder of Nature and Knowledge.* Boston: South End Press, 1997. 148p. ISBN 0-896-08556-2.

A confrontation is shaping up over the legal status and the growing economic importance of genetic diversity and products from underdeveloped countries. Pharmaceutical companies in particular as well as companies with interests in agriculture regard the untapped natural resources and folk medical practices of the developing nations as useful starting points for novel product development. Indigenous peoples are beginning to recognize the real value of what they had been giving away for so many years and are starting to take ownership. The legal status of their proprietary position and the demand for free access to world natural resources for research purposes are the source of much discussion and legal wrangling.

Sibley, Kenneth D., ed. *The Law and Strategy of Biotechnology Patents.* Biotechnology Series No. 25. Boston: Butterworth-Heinemann, 1994. 262p. ISBN 0-750-69444-0.

This book focuses on the law governing biotechnology patents and the strategies followed in protecting intellectual property in the United States.

Weir, B. S., ed. *Human Identification: The Use of DNA Markers.* Contemporary Issues in Genetics and Evolution Series. Boston: Kluwer Academic, 1995. 213p. ISBN 0-792-33520-1.

This monograph explains the methodology and the interpretive issues involved in using DNA patterns to relate individuals.

Wiegele, T. C. *Biotechnology and International Relations: The Political Dimensions.* Gainesville: University of Florida Press, 1991. 212p. ISBN 0-813-01055-1.

The global interdependence in science has created an international political context in biotechnology. This book, written to provide background for those persons pursuing a career in international relations, tries to show where this particular expertise can assist governments in dealing with the new technology to maximize benefits and minimize harm. It reviews national regulatory postures, major international legal instruments, determination of harm and sanctions, and verification and enforcement of laws governing biotechnology applications. Other concerns are the impact of international commerce on both developed and developing economies and the prospect for biological warfare as a terrorist weapon.

World Intellectual Property Organization. *Guide on the Licensing of Biotechnology.* Geneva: World Intellectual Property Organization, Publication No. 708(E), 1992. 197p. ISBN 9-280-50410-X.

The patenting of biotechnology products is a highly specialized branch of intellectual property rights. It is at the center of the commercialization controversy in agriculture, which is far from settled. This book provides a summary of the established concepts that are accepted, at least by developed nations.

Wright, Susan. *Molecular Politics: Developing American and British Regulatory Policy for Genetic Engineering, 1972–1982.* Chicago: University of Chicago Press, 1994. 591p. ISBN 0-226-91065-2.

This book describes the history of the development of regulation of biotechnology and genetic engineering in considerable detail. The author follows the cycle of regulation followed by deregulation with an analysis of the societal, political, economic, and scientific forces that drove events. She interprets the deregulation as a sellout of the public due to overriding industrial (and research) interests rather than to a reasoned change.

Business

Kenney, M. *Biotechnology: The University-Industrial Complex.* New Haven, CT: Yale University Press, 1986. 306p. ISBN 0-300-03392-3.

This book describes the early development of biotechnology and the university-industrial connection from a sociological point of view. The extremely close relationship with industry has not been without its effects on the university. The structure of the biotech industry—venture capital funding, pharmaceutical and chemical company investments, and the rewards and incentives for academics—is placed in perspective. Biotech companies have undergone tremendous growth, which has brought changes in the roles, motivations, and perceptions of their employees. Growing relationships with multinationals are forcing further changes in industry, particularly in those applied to agriculture.

Krimsky, S. *Biotechnics and Society: The Rise of Industrial Genetics.* New York: Praeger. 1991. 280p. ISBN 0-275-93853-X.

A history of the evolution of the industrial involvement in genetic engineering from enhancing technology to commercialization of entirely new ideas and products.

Swanson, T., ed. *Biotechnology, Agriculture, and the Developing World: The Distributional Implications of Technological Change.* Northampton, MA: Edward Elgar, 2002. 281p. ISBN 1-840-64679-9.

This book considers the business effects biotechnology has had and likely will have on modern agriculture and the economies of the developing world. The issues of plant breeders' rights and how genetic engineering and agribusiness are in conflict with agricultural practices in developing countries where much of the farming is by small operations are discussed.

van Balken, J. A. M., ed. *Biotechnological Innovations in Chemical Synthesis.* London: Butterworth-Heinemann, 1997. 376p. ISBN 0-750-60561-8.

This book describes the current and potential uses of biotechnology in chemical synthesis. Industrial fermentation processes are connected with their respective chemical syntheses to compare their strengths and weaknesses and where they might be applied together to advantage.

Agriculture

Beaumont, A. R., and K. Hoare. *Biotechnology and Genetics in Fisheries and Aquaculture.* Oxford: Blackwell Science, 2003. 158p. ISBN 0-632-05515-4.

Most agriculture biotechnology is concerned with plant agriculture. This book presents the impact of genetic engineering on fish and seafood. Increasing numbers of varieties of aquatic and marine organisms are being farm raised for food rather than caught wild. If you eat salmon or catfish at a restaurant, chances are that it is farmed fish. This takes fishing pressure off of those populations but in return presents issues of pollution and questions about practices such as transgenic growth hormone or antibiotic supplementation to support high population densities as well as the consequences of escapes of farmed, genetically altered fish.

*Busch, L., W. B. Lacy, J. Burkehardt, and L. R. Lacy. *Plants, Power, and Profit: Social, Economic, and Ethical Consequences of the New Biotechnologies.* Cambridge, MA: Blackwell. 1991. 275p. ISBN 1-557-86088-2.

This book discusses emerging issues and trends in the biotechnology worldwide. It presents perspectives on science and society including a discussion of the philosophy behind the science. A history of plant breeding and its connection to the new genetic engineering technologies provides a background of the debate about the relative contributions of the two strategies. Examples of political biology with a chapter on wheat and one on tomatoes demonstrate the principles introduced in earlier sections. A chapter on the international scope of biotechnology's social impact demands accountability for releases of transgenic plants and potential economic effects on indigenous industry. There is also the question of who will do the engineering on minority food and other crops important to many developing nations. With the highly technical transgenic crops, the future appears to bode fuller integration of crop production into industry and out of the hands of individual farmers in all countries.

*Dawkins, K. *Gene Wars: The Politics of Biotechnology.* Open Media Pamphlet series. New York: Seven Stories Press, 1997. 60p. ISBN 1-888-36348-7.

This short book makes the case of the perceived perils of plant and animal agricultural biotechnology from the escape of engineered organisms, to the impact on regular agriculture, economic displacement, and the evils of control exercised by the multinational agricultural industry. The points are made succinctly, and the book includes information sources for activists to make a difference.

*Gussow, J. D. *Chicken Little, Tomato Sauce and Agriculture: Who Will Produce Tomorrow's Food?* New York: Bootstrap Press, 1991. 150p. ISBN 0-942-85032-7.

A critical discussion of the impact of genetic engineering on food production and who will be producing it.

Institute of Food Technologies. *Appropriate Oversight for Plants with Inherited Traits for Resistance to Pests.* Chicago: Institute of Food Technologies, 1996.

This report recommends principles for regulating genetically engineered plants. The level of risk should be determined by the characteristics of the plant, not by lack of familiarity with the gene change, the source of genes, or the method of gene transfer. The genes or gene products should not be pesticides that are subject to the federal Insecticide, Fungicide, and Rodenticide Act; concepts such as "generally regarded as safe" (GRAS) applied to other areas such as food additives should also be applied to the new varieties of plants. GRAS substances are not subject to the federal Food, Drug, and Cosmetic Act. This report does not include consideration of global differences of opinion over introducing genetically engineered organisms into agriculture.

Krimsky, S., and R. P. Wrubel. *Agricultural Biotechnology and the Environment: Science, Policy, and Social Issues.* Urbana: University of Illinois Press, 1996. 294p. ISBN 0-252-02164-9.

This book describes the various applications of genetic engineering to agricultural systems. It covers herbicide-resistant plants, insect-resistant plants, disease-resistant crops, various transgenic plant products, microbial pesticides, nitrogen fixation, and frost-inhibiting bacteria. Animal agricultural products are included such as animal growth hormones and their use and various transgenic animals. In keeping with his other publications, Krimsky is careful to include discussion of the cultural and symbolic sides of agricultural biotechnology.

*Mather, R. *Garden of Unearthly Delights: Bioengineering and the Future of Food.* New York: Penguin Books, 1995. 205p. ISBN 0-525-93864-8.

The author, food editor of the *Detroit News*, considers the future of food supply in the United States from the new genetically engineered plants and recombinant bovine growth hormone for meat and milk products to sustainable agriculture through a series of visits with individuals involved in the process. The tone of the commentary and the information selected for presentation are not in favor of the application of biotechnology to agriculture. The author suggests a blueprint for individual action in consumer buying power to support "whole food" production.

Miller, H. I., and G. Conko. *The Frankenfood Myth: How Protest and Politics Threaten the Biotech Revolution.* Westport, CT: Praeger/Greenwood Press, 2004. 296p. ISBN 0-275-97879-6.

The authors present the case that a combination of anti-technology activism, bureaucratic overreach, and business lobbying has resulted in a regulatory framework that is out of sync with the potential benefits and risks of gene splicing. Regulation is accused of stifling new research at both universities and small biotech companies that could resolve some of the issues in contention. They suggest a series of business and public policy reforms that they believe will provide appropriate safeguards while allowing the benefits of advances in agriculture through biotechnology to be available to consumers and farmers. The book is written at the level of upper-level undergraduates but is meant to be read by policy makers as well.

*Nestle, M. *Safe Food: Bacteria, Biotechnology, and Bioterrorism.* Berkeley: University of California Press, 2004. 350p. ISBN 0-520-24223-8.

The author contends that the agribusiness lobby has stifled government regulatory power over foods. Her concern is that this interference occurs for both general foods and genetically modified foods but in different ways. Part 1 covers general food safety and Part 2 covers biotechnology.

Nottingham, S. *Genescapes: The Ecology of Genetic Engineering.* New York: Zed Books/St. Martin's Press, 2002. 211p. ISBN 1-842-77036-5.

This small volume is a reasoned treatment of the ecological impact of the first generation of commercial transgenic plants and animals. The author references multiple studies and is critical of what he characterizes as woefully inadequate field trial studies of the impact of transgenic crops on the ecology, on farming as a livelihood, and on the economic impact in developing countries. He documents the continuation of the process of industrialization of agriculture begun during the Green Revolution and the tightening control of multinational corporations over the food supply. Many of the same issues of plant monoculture effects in the surrounding ecology are presented with potentially farther reaching consequences.

Although not a diatribe against genetic engineering, the author clearly feels that the multinational corporations that dominate the area have a different set of priorities than societies do. He believes that governments should act to balance the economic forces with the social and ecological consequences of the application of new technology.

*Perlas, N. *Overcoming Illusions about Biotechnology.* Atlantic Highlands, NJ: Zed Books, 1994. 119p. ISBN 1-856-49303-2.

The author considers the impact of genetic engineering and biotechnology on the environment, natural resources, society, and the family farm to be largely negative. He describes sustainable agricultural methods as a solution and the control of biotechnology by legislative and populist means.

Pringle, P. *Food, Inc. Mendel to Monsanto — The Promise and Perils of the Biotech Harvest.* New York: Simon and Schuster, 2003. 239p. ISBN 0-743-22611-9.

A revealing analysis of the propaganda battles and fear mongering by technologists and agribusiness and by anticorporate eco-warriors over genetically modified foods. The author is a journalist who provides an evenhanded and entertaining account of this ongoing dispute.

Reborn, P. *The Last Harvest: The Genetic Gamble That Threatens to Destroy American Agriculture.* Lincoln: University of Nebraska Press, 1995. 269p. ISBN 0-803-28962-6.

The loss of genetic diversity brought about by current plant breeding programs and the cultivation of vast monocultures is being exacerbated by the closing of seed germ plasm repositories. The author points out the vulnerability of homogeneous crop plants to disease and environment without periodic access to the heterogeneity of traits offered by other varieties freely exchanging genetic material in the wild.

Rissler, J., and M. Mellon *The Ecological Risks of Engineered Crops.* Cambridge, MA: MIT Press, 1996. 168p. ISBN 0-26218-171-1.

This monograph sponsored by the Union of Concerned Scientists points out potential risks associated with the widespread commercial cultivation of transgenic crop plants. Most current traits being modified are specific pest and herbicide resistance and traits that benefit processors but don't enhance nutritional value. The original idea of making plants drought- or salt-resistant, nitrogen fixing, or higher yielding has faded with the realization that these are multigenic traits that are less amenable to transgenic manipulation. The authors propose a specific testing program to determine the risks of a given transgenic plant becoming a weed or passing the transgene to wild populations.

Environment

Chaudhry, G. R., ed. *Biological Degradation and Bioremediation of Toxic Chemicals.* Portland, OR: Dioscorides Press, 1994. 515p. ISBN 0-931-14627-5.

This book describes the vast microbial metabolic diversity and the quest to construct recombinant strains to combine pathways in the same organism to improve their effectiveness in bioremediation.

Kalaitzandonakes, N. *Economic and Environmental Impacts of Agbiotech: A Global Perspective.* New York: Kluwer/Plenum, 2003. 336p. ISBN 0-306-47501-4.

An international analysis of the growing influence of agribusiness and, in particular, genetic modifications of organisms on world food supply and the environment.

Prieels, A.-M. *Development of an Environmental Bio-Industry: European Perceptions and Prospects.* Lanham, MD: Dublin Institute, by the European Foundation for the Improvement of Living and Working Conditions. Distributed by UNIPUB, 1993. 131p. ISBN 9-282-64691-2.

This report is compiled from responses (in 1990) of persons and organizations to two general questions: can biotech contribute to an improvement in the state of the environment, and why is bio-industry not more developed in Europe? Specific topics include the pollution treatment bio-industry, the scientific and technological bases of the environmental bio-industry, acceptance of the development of applications in the environmental sector, biotechnology and sustainable development, and the political and regulatory framework of the environmental bioindustry. Many of the problems are the same as those encountered in the United States.

Thomas, J. A., and R. L. Fuchs, eds. *Biotechnology and Safety Assessment.* San Diego, CA: Academic Press, 2002. 487p. ISBN 0-126-88721-7.

How exactly does one go about demonstrating the safety of biotechnology and how does it compare with chemical safety? What is safe enough and who sets the standards?

Witt, S. C. *Biotechnology, Microbes, and the Environment.* Brief-book series. San Francisco: Center for Science Information, 1990. 219p. ISBN 0-912-00503-3.

This book is designed to provide decision makers with a basic level of scientific understanding to make informed public policy; it is also intended to educate the general public. Written in a distinctly irreverent but engaging style, the majority of the book is concerned with the impact of the technology, risk assessment, U.S. and international regulation, historical events, and major issues at stake. A unique contribution is a list of U.S. and international expert sources drawn from academic scientists, government agency workers, industrial representatives, and foes of genetic engineering with comments on their areas of expertise and an assessment of the "articulate interview" and the "solid source."

Science

Bronzino, J. E., ed. *The Biomedical Engineering Handbook.* Boca Raton, FL: CRC Press, 1995. 2,862p. ISBN 0-849-38346-3.

A compendium on the design, development, and use of medical technology to diagnose and treat patients. It includes 11 chapters on biotechnology and 17 chapters on tissue engineering. Regulations and ethics currently in force that influence biomedical engineering are discussed.

Claverie, J.-M., and C. Notredame. *Bioinformatics for Dummies.* Foster City, CA: IDG Books Worldwide, 2003. 432p. ISBN 0-764-51696-5.

Written in the breezy style of the For Dummies series, this book provides a quick introduction to bioinformatics and how to use it. Although a lot of information is packed into a small space, the format allows the reader to appreciate and use the extensive resources of the World Wide Web to do research or just to learn about history, techniques, and new developments—all from a desktop computer.

Danchin, A. *The Delphic Boat: What Genomes Tell Us.* Cambridge, MA: Harvard University Press, 2002. 368p. ISBN 0-674-00930-4.

In explaining what a genome is and what it codes for, the author's emphasis is as much on the arrangement and spacing of the coding components in the nucleotide sequence as on what the protein and known regulatory sequences are. Genome science is at the point where it has determined the linear sequence of the genetic message, but having that information does not allow the creation of a new cell, much less a whole organism. Something else is needed, an organizational plan. The author suggests that this plan may be encoded in some way within existing organisms building off of a preexisting template. His interpretation is that by the time in evolutionary history that genomes had assembled, the genetic code may not be able to stand on its own. According to this scenario, a dinosaur could not be produced from its DNA.

Friedmann, T. *Gene Therapy: Fact and Fiction in Biology's New Approaches to Disease.* Plainview, NY: Cold Spring Harbor Laboratory Press, 1994. 124p. ISBN 0-879-69446-7.

This short book provides historical background and an assessment of current challenges. It reviews the field's technical achievements and ethical dilemmas.

Gralla, J. D., and P. Gralla. *Complete Idiot's Guide to Understanding Cloning.* New York: Alpha Books, 2004. 323p. ISBN 1-592-57148-4.

One of the highly readable Complete Idiot series of overview books that provide an introduction to a wide range of topics. There's a lot of information on both the science and the social impact of nuclear cloning technology. After several introductory chapters that give history and explanation of genetic engineering principles, a large fraction of the book is devoted to nuclear cloning and stem cells, which have engendered the most public controversy although they are the least frequent applications of genetic engineering technology. There is also coverage of gene therapy with its high hopes and thus far meager clinical success.

Hall, S. S. *Invisible Frontiers: The Race to Synthesize a Human Gene.* New York: Atlantic Monthly Press, 1987. 334p. ISBN 0-871-13147-1.

Story of the race to clone and express human insulin for the treatment of diabetes.

Jackson, J. F. *Genetics and You.* Totowa, NJ: Humana Press, 1996. 92p. ISBN 0-896-03329-5.

Written for the educated lay consumer of genetic testing, this short book explains the principles of genetics and how to make use of the test results in conjunction with genetic counseling. It clarifies the new genetics and provides guidance on how to get questions answered and deal with the decisions to be made.

Kreuzer, H., and A. Massey. *Recombinant DNA and Biotechnology: A Guide for Teachers;* and *Recombinant DNA and Biotechnology: A Guide for Students.* American Society for Microbiology Press, 1996. 564p. ISBN 1-555-81101-9C (teacher); 364p. ISBN 1-555-81110-8C (student).

An introductory textbook for high school and college students containing sections of textbook topics, wet lab experiments, dry lab experiments and demos, and material on weighing the risk-benefit impact on society. The book includes a decision-making model for bioethical issues, case studies in bioethics, and information on careers in biotechnology.

Miklos, D. A., and G. A. Freyer. *DNA Science: A First Course in Recombinant DNA Technology.* Plainview, NY: Cold Spring Harbor Laboratory Press, 1990. 477p. ISBN 0-892-78411-3.

This is a combination textbook and lab manual designed for advanced high school or beginning college students showing where recombinant DNA technology came from and pointing to where it might lead. It traces a historical perspective with experiments designed to illustrate the text. The course is designed to be supported by Carolina Biological Supply reagents and supplies if required. Well illustrated and engagingly written, the textbook itself is suitable for introducing recombinant DNA in courses on science and society.

Moody, G. *Digital Code of Life: How Bioinformatics Is Revolutionizing Science, Medicine, and Business.* New York: John Wiley and Sons, 2004. 400p. ISBN 0-471-32788-3.

Describes how the field of bioinformatics came about and profiles the major players in the field, their personalities, and their contributions. The first important contribution of the new discipline of bioinformatics was as an indispensable element in the piecing together of thousands of nucleotide strings in the proper order to give the complete human genome. New information technology also was key to analyzing the billions of nucleotides in the sequence to extract information encoded there. The book explores the commercial applications and investment opportunities as well as when and where future growth may occur.

Murray, T. H., M. A. Rothstein, and R. F. Murray, Jr. *The Human Genome Project and the Future of Health Care.* Bloomington: University of Indiana Press, 1996. 248p. ISBN 0-253-33213-3.

This book focuses on the implications of the Human Genome Project for society and health practice. Attention is drawn to the effect the information from the Genome Project will have on the quality and delivery of healthcare in the United States. The issues of genetic discrimination in employment and insurance and equal access to testing and treatment of minority and indigent clients as well as the impact on reproductive decision making are treated in a series of articles by experts in those areas.

Neel, J. V. *Physician to the Gene Pool: Genetic Lessons and Other Stories.* New York: John Wiley and Sons, 1994. 457p. ISBN 0-471-30844-7.

The author, a physician, discusses genetics and populations, genetics and individuals, and the impact of medical genetics on past and present medical care. He then compares these effects to those expected from the application of genetic engineering.

Prentice, D. A. *Stem Cells and Cloning.* San Francisco: Benjamin Cummings, 2003. 39 pp. ISBN 0-805-34864-6.

A short, concise introduction to embryonic stem cells and nuclear cloning technologies.

Rabinow, P. *Making PCR: A Story of Biotechnology.* Chicago: University of Chicago Press, 1996. 190p. ISBN 0-226-70146-8.

An entertaining story of the development of the polymerase chain reaction (PCR) method, a key technology in genetic engineering. The author provides a historical description of the events during the conception of this powerful tool for manipulating nucleic acids. It engagingly includes numerous personal accounts of the struggle for power and precedence among the key players.

Svendsen, P., and J. Hau, eds. *Handbook of Laboratory Animal Science. Volume 1: Animal Models.* Boca Raton, FL: CRC Press, 1994. 224p. ISBN 0-849-34378-X. *Handbook of Laboratory Animal Science. Volume 2: Selection and Handling of Animals in Biomedical Research.* Boca Raton, FL: CRC Press, 1994. 448p. ISBN 0-849-34390-9.

A comprehensive description of lab animal genetics, diseases, health monitoring, nutrition, and environmental impact on animal testing. It considers the ethics of animal experimentation in Europe and North America. Also discussed are alternatives to animal experiments, including isolated organs, cell cultures, and computer simulations.

Books on Genetic Engineering (Young Adult Sources)

Balkwill, F. *Amazing Schemes within Your Genes.* London: HarperCollins, 1993. 32p. ISBN 0-001-96465-8.

The uniqueness of every human being is explained through their genes, even though our DNA is 99.5% identical. The process by which visible characteristics such as hair, eye, and skin color or ear shape are transmitted and how genetic diseases such as cystic fibrosis are inherited are explained and illustrated for readers aged 9 to 15 years.

Balkwill, F. *DNA Is Here to Stay.* London: HarperCollins, 1992. 32p. ISBN 0-001-91165-1.

The process by which the code of life directs the growth from an embryonic cell to a complete human being is described and clearly illustrated for readers aged 9 to 15 years.

Bornstein, S. *What Makes You What You Are: A First Look at Genetics.* Englewood Cliffs, NJ: J. Messner, 1989. 128p. ISBN 0-67168-650-X.

An introduction to genes, DNA, and genetics for grades 7 and up.

Bryan, J. *Genetic Engineering.* New York: Thomson Learning, 1995. 64p. ISBN 1-568-47268-4.

The main topic of this book is the issues involved in the application of genetic engineering. The treatment is up-to-date as of 1995, and the illustrations and pictures are very good.

Darling, D. J. *Genetic Engineering: Redrawing the Blueprint of Life.* Parsippany, NJ: Silver Barnett, Dillon Press, 1995. 64p. ISBN 0-875-18614-9.

This book outlines the progress made in understanding the genetic code by following research on cystic fibrosis. The author uses this disease with major effects on children to explain mechanisms of genetic diseases and to enhance student involvement with the issues of genetic screening, amniocentesis, abortion, and gene therapy. For grades 5 and up.

Gerbi, S. A. "From Genes to Proteins." In **Carolina Biology Readers Series, Volume 158.** Burlington, NC: Carolina Biological Supply. 16p. ISBN 0-892-78358-3.

A description of the biochemical events involved in information flow from a gene sequence to the functioning protein. For grades 10–12.

Grace, E. S. *Biotechnology Unzipped: Promises and Realities.* Washington, DC: Joseph Henry Press (National Academy Press), 1997. 264p. ISBN 0-309-05777-9.

An up-to-date discussion of the possibilities and achievements of the new science of biotechnology.

Lampton, C. *DNA and the Creation of New Life.* New York: ARCO Publishing, 1983. 135p. ISBN 0-668-05396-8.

This somewhat dated book details the story of the quest for the gene and what scientists are doing to mold genes to do their bidding. The author, a well-known contributor to juvenile science writing, also considers some of the issues scientists face as the "engineers of life," including genes as property. Discussion of both the projected uses and issues of DNA technology is similar to those current now.

Lampton, C. *DNA Fingerprinting.* Danbury, CT: Franklin Watts, 1991. 112p. ISBN 0-531-13003-7.

An explanation of how DNA patterns in cells and body fluids are much like a fingerprint and can be used to identify individuals. DNA analysis evidence can be used in a court of law to place an individual at the scene of a crime. For grades 7 to 12.

Marshall, E. L. *The Human Genome Project: Cracking the Code Within Us.* Danbury, CT: Franklin Watts, 1996. 128p. ISBN 0-531-11299-3.

The fifteen-year, multibillion-dollar project of mapping and sequencing the human genetic code, the genome, is described and explained in this work for grades 8 to 12. The author explains the lay concepts involved in genetic research by concentrating on individual real scientists and the work they do as part of the project. The controversies involved in doing and applying genetic research are recognized, along with the potential benefits of the knowledge for research and human health, brought to life again through the device of real-life stories about the people involved. Personalizing in this way helps to drive home the magnitude of the Human Genome Project. It also serves as a reminder that in the end it all comes down to ordinary individuals—the public—who will have to decide how to use the information.

Marteau, T., and M. Richards, eds. *The Troubled Helix: Social and Psychological Implications of the New Human Genetics.* Cambridge: Cambridge University Press, 1996. 359p. ISBN 0-521-46288-6.

This book addresses the interpretation and use of genetic information at both the personal level and by society as a whole. As the authors, both pro and con, of the various chapters on carrier testing of adults, prenatal testing, testing children, and the impact of genetic counseling point out, disagreement rests on the lack of predictability of the outcome of having a genetic defect. Although the technology for determining that there is a mutation in a specific gene has improved immensely, being able to predict when, where, and how much of an effect that lesion will have on an individual has not progressed appreciably. The interpretation problem will only get worse as geneticists tie together more complex multigenic traits.

The *Troubled Helix* also analyzes social aspects of genetic testing. How do people react to and use genetic information to make decisions? Many people who are at risk for a devastating disease simply do not wish to know. Other points of debate include testing of children and how and when to inform other kindred of test results. There is public agreement throughout that genetic information is private for the individual. Numerous studies and statistics of various types are cited. The book is written from the English point of view, although a great deal of effort is made to make things more international. The reader will note that certain basic assumptions about public attitudes and legal precedence differ from those in the United States.

Sherrow, V. *James Watson and Francis Crick: Decoding the Secret of DNA.* Woodbridge, CT: Blackbirch Press, 1995. 110p. ISBN 1-56711-133-5.

This historical-biographical account of the discovery of the structure of DNA is written for the middle school and early high school student and contains numerous illustrations and photographs of the important principles and human participants in the events leading up to the determination of the three-dimensional structure of the DNA molecule. The story of Watson and Crick and DNA is carried up through the 1990s, including Watson's brief tenure as head of the Human Genome Project.

Swisher, C. *Genetic Engineering.* San Diego, CA: Lucent Books, 1996. 128p. ISBN 1-560-06179-0.

This book provides a comprehensive overview of genetic engineering technology placed in the perspective of the controversy brought by its applications. Opposing points of view are presented with equal weight to the proponents. A glossary, extensive index, list of organizations to contact for additional information, and suggestions for further reading make this book a good resource for young readers and some adults. For grades 7 and up.

Tagliaferro, L. *Genetic Engineering: Progress or Peril?* Minneapolis, MN: Lerner, 1996. 128p. ISBN 0-822-52610-7.

The author balances the potential benefits of human, plant, and animal genetic experimentation with cautionary statements. Separate chapters on each of these topics follow a short introduction on cell and DNA structure and function and descriptions of genetics and Mendelian heredity theory. Discussion of human gene identification, the development of new life forms, and the regulation of genetic engineering lead into considerations of patent usage and the effects of exclusive use of genetic modifications of current organisms turning into monopoly. A substantial bibliography including articles from periodicals, corporate scientific reports, and monographs that are a rich source for debating current issues in genetic engineering is provided. For grades 7 to 9.

Thro, E. *Genetic Engineering.* New York: Facts on File, 1993. 128p. ISBN 0-816-02629-7.

Through the use of charts and line drawings, the author covers the usual topics of cell and DNA structure, chromosomes, and genes. She outlines some uses that have already been found (as of 1993) for genetic engineering and warns of upcoming ethical problems in eugenics and patenting of new life forms. For grades 7 to 12.

Van Loon, B. *DNA, The Marvelous Molecule.* Norfolk, England: Tarquin, 1990. 32p. ISBN 0-906-21275-8.

The helical structure of DNA and the topology of the many structures that inhabit the microscopic world of molecular biology are often inadequately portrayed on the flat page. The three-dimensional heavy paper models provided by Van Loon to be cut out and assembled solve this problem. The models include a DNA helix, a bacteriophage with a packaged DNA helix, nucleotide hydrogen bonding pairs, and a protein-folding model that graphically illustrates the effect of a mutation on a protein's shape. A 27-page minibook included within the package succinctly and clearly explains the models and the place of DNA in life, genetics, and evolution. This book is written for middle and senior high school students.

Wekesser, C., ed. *Genetic Engineering: Opposing Viewpoints.* San Diego, CA: Greenhaven Press, 1997. 240p. ISBN 1-565-10358-0.

A discussion of the benefits and risks of the application of genetic engineering technology to medicine, agriculture, industry, and the environment. Ethical issues such as human cloning and genetic testing are also covered. For grades 5 to 12.

Wells, D. K. *Biotechnology.* Tarrytown, NY: Benchmark Books (Marshall Cavendish), 1997. 64p. ISBN 0-761-40046-X.

A description of the tools and wonders of biotechnology. For grades 3 to 5.

U.S. Government Publications

Office of Technology Assessment Documents

Titles are available from:

U. S. Government Printing Office National Technical
Information Service
Superintendent of Documents
5285 Port Royal Road
Department 33, Springfield, VA 22161-0001
Washington, DC 20402-9325 (703) 487-4650
(202) 783-3238

Impacts of Applied Genetics: Micro-Organisms, Plants, and Animals

OTA-HR-132, April 1981.

An Assessment of Alternatives for a National Computerized Criminal History System

OTA-CIT-161, October 1982.

Splicing Life: A Report on the Social and Ethical Issues of Genetic Engineering with Human Beings

Pr 40.8:Et 3/L 62. November, 1982.

Commercial Biotechnology: An International Analysis

OTA-BA-218, January 1984.

Human Gene Therapy: A Background Paper

OTA-BP-BA-32, December 1984.

Technologies for Detecting Heritable Mutations in Human Beings

OTA-H-298, September 1986.

Ownership of Human Tissues and Cells—Special Report

OTA-BA-337, March 1987.

New Developments in Biotechnology: Public Perceptions of Biotechnology

OTA-BP-BA-45, May 1987.

Issues Relevant to NCIC (National Crime Information Center) 2000 Proposals

Staff paper, November 1987.

Mapping Our Genes: Genome Projects—How Big? How Fast?

OTA-BA-373, April 1988.

Field Testing Engineered Organisms: Genetic and Ecological Issues—Special Report

OTA-BA-350, May 1988.

U.S. Investment in Biotechnology: An International Analysis

OTA-BA-360, July 1988.

Medical Testing and Health Insurance

OTA-H-384, August 1988.

Patenting Life—Special Report

OTA-BA-370, April 1989.

Genetic Witness: Forensic Uses of DNA Tests

OTA-BA-483, July 1990.

Genetic Monitoring and Screening in the Workplace: Contractor Documents

OTA-BA-455, October 1990.

Genetic Screening in the Workplace
OTA-BA-456, October 1990.

Biotechnology in a Global Economy
OTA-BA-494, October 1991.

Cystic Fibrosis and DNA Tests: Implications of Carrier Screening
OTA-BA-532, August 1992.

Genetic Counseling and Cystic Fibrosis Carrier Screening: Results of a Survey
OTA-BP-BA-97, September 1992.

Genetic Counseling and Cystic Fibrosis Carrier Screening: Results of a Survey (background paper)
OTA-BP-BA-98, September 1992.

Biomedical Ethics and U.S. Public Policy. Hearing of the Committee on Labor and Human Resources
OTA-BP-BBS-105, June 1993.

Congressional Reports

Biotechnology and the Ethics of Cloning: How Far Should We Go? Hearing before the Committee on Science, Subcommittee on Technology, U.S. Congress. 1997. 59p. ISBN 0-160-55267-2.

Biotechnology Science Competitiveness Act of 1988. U.S. House Committee on Agriculture. CIS 88 #703-15 (pt. 1); CIS 88 #163-18 (pt. 2).

A Coordinated Framework for the Regulation of Biotechnology. *Fed. Reg.* 51: 23303-23309 (1986).

Designing Genetic Information Policy: The Need for an Independent Policy Review of the Ethical, Legal, and Social Implications of the Human Genome Project (16th report). (No. #102-478 USGPO), House Committee on Government Operation, U.S. Congress. 1992.

Genetic Non-discrimination: Implications for Employers and Employees. Hearing before the Subcommittee on Employer-Employee Relations of the Committee on Education and the Workforce. 2002. 154 pp. U.S. Government document number Y4.ED8/1:107-25.

The Genome Project: The Ethical Issues of Gene Patenting. Hearing before the Committee on the Judiciary, Subcommittee on Patents, Copyrights, and Trademarks. 1993. 240p. ISBN 0-160-41610-8.

Guidelines for Research Involving Recombinant DNA Molecules. *Fed. Reg.* 59: 34496-34547 (1994).

Issues in the Federal Regulation of Biotechnology: From Research to Release. By the U.S. Congress House Committee on Science and Technology, Subcommittee on Investigations and Oversight. CIS 86 #702-18, December 1986. 118 pp.

Other Government Agency Reports

Biotechnology Law for the 1990's: Analysis and Perspective. By the Special Projects Unit of the Bureau of National Affairs, the BNA Special Report Series on Biotechnology; Special Report #4, Bureau of National Affairs, Washington, DC. 1989. ISBN 1-558-71153-8.

Blair, R. R. Forensic DNA Analysis: Issues. Washington, DC: U.S. Department of Justice, Office of Justice Programs, Bureau of Justice Statistics. 1991. 32p. GPO item # 0968-H-12 Doc. J 29.9/8:F 76.

Committee on Health, Education, Labor and Pensions. Genetic Information in the Workplace. Washington, DC, 2000. 121p. Government document number Y4.L 11/4 : S.HRG.106-647.

Recombinant DNA Research, Volume 20: Documents relating to "NIH Guidelines for Research Involving Recombinant DNA Molecules," August 1994 to December 1994, NIH Publication number 95-3993, U.S. Department of Health and Human Services, Public Health Service, NIH Report of Recombinant Advisory Committee, December 1995.

Report on National Biotechnology Policy, the President's Council on Competitiveness. Government document number Pr 41.8:B52 1991.

U.S. Biotechnology: A Legislative and Regulatory Roadmap. By the Special Projects Unit of the Bureau of National Affairs, the BNA Special Report Series on Biotechnology; Special Report #2, Bureau of National Affairs, Washington, DC. 1989. ISBN 1-558-71139-2.

Nongovernment Reports

President's Council on Bioethics. **Beyond Therapy: Biotechnology and the Pursuit of Happiness. A Report by the President's Council on Bioethics.** New York: Regan Books, HarperCollins, 2003. 328p. ISBN 0-060-73490-6.

This report divides a discussion of the ethical implications for aspects of biotechnology in four areas: Better Children, Superior Performance, Ageless Bodies, and Happy Souls. Issues that they address are humility, unnaturalism, individuality, access, unfairness, liberty, health, commerce, medicine, and American ideals.

The Evaluation of Forensic DNA Evidence: An Update. National Research Council, Committee on DNA Forensic Science. Washington, DC: National Academy Press, 1996. 254p. ISBN 0-309-05395-1.

Genetic Information and Insuring: Confidentiality Concerns and Recommendations, American Council of Life Insurance, Subcommittee on Privacy Legislation. 1990.

Genetic Tests and Health Insurance: Results of a Survey. Upland, PA: Dione, 1993. ISBN 1-56806-637-6.

Miller, T. H. **The Human Genome Project and Genetic Testing: Ethical Implications.** Series: The Genome, Ethics, and the Law: Issues in Genetic Testing, Washington, DC: American Association for the Advancement of Science, 1991.

van Overwalle, G. **Opinion on Ethical Aspects of Patenting Inventions Involving Human Stem Cells. Opinion No. 16.** Luxembourg: European Group on Ethics in Science and New Technologies to the European Commission, 2002. 135p.

Report of the ACLI-HIAA Task Force on Genetic Testing. Washington, DC: Health Insurance Association of America, 1991.

van Overwalle, G. **Study on the Patenting of Inventions Related to Human Stem Cell Research.** Luxembourg: European Group on Ethics in Science and New Technologies to the European Commission, 2002. 218p. KA-41-01-470-EN-C.

Periodicals and Newsletters

The Ag Bioethics Forum
115 Morrill Hall, Iowa State University
Ames, IA 50011

Interdisciplinary coverage of agricultural bioethics.

Applied Genetic News
Business Communications Company, Inc.
25 Van Zant St.
Norwalk, CT 06855-1781
Online vendor: DIALOG (Knight-Ridder Information, Inc.)
http://www.buscom.com
Monthly. $395/yr

This newsletter covers the application of genetic research to industry and technology and evaluates ongoing research in aging as well as cancer and other diseases. Research funding, venture capital, and stock prices are also discussed. First published in 1980, the targeted audience is both biotechnology professionals and laypersons.

BIO News (Biotechnology Industry Organization)
Biotechnology Industry Organization
1625 K Street, NW, Suite 1100
Washington, DC 20006
Six/yr. Free to members

This newsletter concentrates on federal regulations and legislative developments affecting the biotechnology industry. Since 1985, it has served as the Association of Biotechnology Companies newsletter.

BioTech Market News & Strategies
Conmar Enterprises, Inc.
P.O. Box 11155
Ft. Lauderdale, FL 33339
Monthly. $322/yr

First published in 1983, this newsletter contains marketing and product development strategies to biotechnology and pharmaceutical executives. In addition it covers "new marketable applications" and summarizes lawsuits related to biotechnology, patents, and environmental regulations.

Biotechnology and Development Monitor. University of Amsterdam, the Netherlands: (joint) Directorate General, International Cooperation of the Ministry of Foreign Affairs, the Hague, Switzerland, and the University of Amsterdam, the Netherlands.

International monitoring of biotechnology.

Biotechnology Law Report
Mary Ann Liebert, Inc.
20 W. 3rd Street, 2nd Floor
Media, PA 19063-2824
Bimonthly. $605/yr

This hefty newsletter, up to 200 pages per issue, covers the legal developments in the fields of biotechnology and genetic engineering. First published in 1982, it includes aspects of product liability, patent, biomedical, contract and licensing, and international law along with pertinent legislation, regulatory actions, litigation resolution, and international developments. It publishes complete texts of significant court decisions, briefs, regulations, and legislation. Aimed primarily at lawyers with biotechnology clients, regulatory affairs professionals, and university and company biotechnology research departments.

Editor: Gerry J. Elman (chief)

Biotechnology News
CTB International Publishing, Inc.
P. O. Box 218
Maplewood, NJ 07040-0218
Thirty/yr; indexed semiannually. $538/yr

This newsletter concentrates on developments that affect genetic engineering, microbial and enzyme technology, and fermentation contributions to the production of pharmaceuticals, foods, crops, fuels, and chemicals. Established in 1980, this publication is mostly targeted at biotechnology industry management.

Biotechnology Newswatch
The McGraw-Hill Companies
1221 Avenue of the Americas, 36th Floor
New York, NY 10020
Semimonthly. $975/yr

This newsletter is intended to provide an overview of the international biotechnology industry through a series of capsule summaries of news stories and a compilation of research and business stories. It is aimed at biotechnology industry professionals.

Biotechnology: The International Monthly for Industrial Biotechnology
65 Bleeker Street
New York, NY 10012-2467
Nature Publishing Company

Biotech Reporter
Freiberg Publishing
P. O. Box 7
Cedar Falls, IA 50613
Monthly. $135/yr

Formerly *AgBiotechnology News* and published since 1984, the *Reporter* covers business and technological aspects of international agricultural biotechnology. Besides the industrial perspective, it also covers educational opportunities related to the field.

BLAST, Bulletin of Law, Science & Technology
Section of Science and Technology—American Bar Association (ABA)
750 N. Lake Shore Drive
Chicago, IL 60611
Quarterly. Included in ABA membership

This professional newsletter of the American Bar Association, established in 1976 as the Bulletin of Law, Science & Technology, is concerned with current issues relating law and science and technology. Subjects included are the use of computers and the law, controls on scientific information, legal problems of genetic technologies, and policy issues in the communications industry.

Carolina Tips
Carolina Biological Supply Company
2700 York Road
Burlington, NC 27215
Quarterly; indexed annually. Free to science teachers on written request on school letterhead and to health professionals

This newsletter contains articles of interest to science teachers from the elementary to college levels. Published by Carolina Biological Supply, a major source for science teaching supplies since 1938, it contains many useful tips for teaching, well-illustrated articles, and student experiments.

Genetic Engineering News
New York: Mary Ann Liebert
Monthly.

Current news on science and industry developments in genetic engineering technology. Includes business information, new companies, new products. Commentary and articles on issues of interest to the genetic engineering and biotechnology community.

The Gene Exchange: A Public Voice on Genetic Engineering
National Biotechnology Policy Center

National Wildlife Federation
1400 16th Street, NW
Washington, DC 20036

Genetic Engineering Letter
8750 Georgia Avenue, Suite 124
Silver Spring, MD 20910

GeneWatch
19 Garden Street
Cambridge, MA 02138

Newsletter of the Council for Responsible Genetics, a resource for public involvement.

Human Genome News
Human Genome Management Information System for the U.S. Department of Energy
Oak Ridge National Laboratory
1060 Commerce Park
Oak Ridge, TN 37830
Quarterly. Free

Formerly the *Human Genome Quarterly* and first published in 1989, this governmental newsletter specializes in news about the Human Genome Project and is intended to facilitate communications among genome researchers. Besides the researchers, consumers of human genome information such as teachers, genetic counselors and physicians, ethicists, students, congressional staff, and the general public may find the information useful.

Intellectual Property & Biodiversity News
Institute for Agriculture and Trade Policy
2105 1st Avenue, South
Minneapolis, MN 55404
Unknown. Free

Aimed at biotechnologists, this newsletter discusses the current news headlines that affect biotechnology.

Life Sciences & Biotechnology Update
Infoteam, Inc.
P. O. Box 15640

Plantation, FL 33318-5640
Monthly. $269/yr

Aimed at professionals involved with the applications of biotechnology in the life sciences, this newsletter covers activities in agriculture, food, medicine, health, biological and biomedical engineering, microorganisms, nutrition, and disease.

McGraw-Hill's Biotechnology Newswatch
McGraw-Hill, Inc.
1221 Avenue of the Americas, 36th Floor
New York, NY 10020
Online vendor: LEXIS-NEXIS; DIALOG (Knight-Ridder Information, Inc.); Dow Jones
Semimonthly. $875/yr

A comprehensive update since 1981 of news of scientific, commercial, and governmental significance in the biotechnology field. Areas included are genetic engineering, hybridoma technology, applied plant genetics, enzymology, and biomass conversion. It is aimed at a wide audience interested in the progress and future of biotechnology, from scientists to investment analysts, patent attorneys, and government officials.

Mealey's Litigation Report: Biotechnology
Mealey Publications, Inc.
512 W. Lancaster Avenue
P.O. Box 446
Wayne, PA 19087-0446
Semimonthly; indexed annually. $650/yr

One of several newsletters specializing in various topics of legal litigation, the *Report,* beginning in 1996, has covered in depth the legal disputes over processes and products developed through the use of biotechnology "from DNA to drugs to disease detection." Legislative action and international disputes as well as expert commentary are included. It is directed primarily to legal counsel for biotechnology, chemical, and pharmaceutical companies.

New England Regional Genetics Group—Regional Newsletter
New England Regional Genetics Group
P.O. Box 670

Mt. Desert, ME 04660
Semiannually. Free

First published in 1985, this newsletter covers news of interest to genetic service professionals and consumers of genetic services in the New England region.

Olsen's Biotechnology Report
G. V. Olsen Associates
123 Picketts Ridge Road
West Redding, CT 06896
Monthly. $150/yr

Focused primarily on food and agricultural biotechnology, this newsletter provides coverage on genetic engineering research, scale-up, production, and marketing of plant, animal, and agricultural chemicals. It also includes a financial analysis of the biotechnology market and investment opportunities in that area. First published in 1983, this newsletter lists large farmers and growers as well as universities as a part of its audience.

U.S. Regulatory Reporter
Parexel International Corporation
195 West Street
Waltham, MA 02154-1116
Monthly. $395/yr

This newsletter, first published in 1984, covers the regulatory news from the Food and Drug Administration primarily for pharmaceutical and biotechnology industry professionals. It provides analysis of FDA product approval standards and the review process.

West Coast Biowatch
2700 H Street
Sacramento, CA 95816

Your World: Biotechnology and You
State College, PA: Pennsylvania Biotechnology Association.

A journal for junior and senior high students describes the application of biotechnology to problems facing our world. Science and engineering.

Directories

Woolum, J., ed. *AgriBioScan.* Phoenix, AZ: Oryx Press, 320p. *Three times/yr. $395/yr.*

A compilation of statistics on agricultural biotechnology companies. Information includes company name, address, phone, fax, research and development, key personnel, financial data, partnerships, and products awaiting Food and Drug Administration or U.S. Department of Agriculture approval.

Walters, L., T. J. Kuhn, and D. M. Goldstein, editors. *Bibliography of Bioethics.* Washington, DC: Kennedy Institute of Ethics, 2005. 701 p. ISBN 1-883-91312-8.

The bibliography covers the whole of bioethics. The topics relevant to genetic engineering are cloning (pp. 97–102); gene therapy (pp. 167–171); genetic counseling (pp. 177–188); and recombinant DNA research (pp. 334-335).

Brogna, C., ed. *Bioindustry Directory.* Maplewood, NJ: CTB International. 165p. *Annual. $157/yr.*

A listing of more than 3,000 U.S. and foreign companies, government agencies, cultural collections, professional organizations, and trade groups working in genetic engineering, plant biotechnology, applied molecular biology, and bioculture technologies. Information included: Name of firm or agency, address, phone, fax.

Coombs, J., and Y. R. Alston, ed. *Biotechnology Directory.* New York: Stockton Press, 500p. *Annual, December. $275 + shipping and handling*

Lists more than 10,000 companies, universities, research centers, government agencies, and suppliers of products and services. Information includes: organization name, address, phone, fax, contact, description of products, services, or research.

Dibner, M. D. *Biotechnology Guide USA: Companies, Data, and Analysis.* New York: Macmillan Publishers in association with the Institute for Biotechnology Information, 1999. 710 pp. ISBN 0-333-79409-5. *Irregular publication frequency.*

This book is also available as an electronic database: U.S. companies database—1,100 companies; actions database—more than 10,000 strategic actions and alliances in commercial biotechnology from 1981; other databases on general topics of interest to commercial biotechnology. Information includes address, phone, fax, financing, names and titles of key management personnel, products on the market and in R&D, number of employees, technologies used, partnerships, and failed companies.

Han Consultants. *Chinese Biotechnology Directory.* Wuhan, Hubei, People's Republic of China: Han Consultants, 1993. 256p. ISBN 7-501-10570-9.

Biotechnology in the People's Republic of China; claims to be the only collected source in English on biotechnology in mainland China. Information includes lists of companies, location, products, areas of interest, and contacts. It also provides an overview of biotechnology in mainland China; a description of government policy on technology transfer, intellectual property, and patent coverage; and a list of information resources in China— abstracting services, journals, and newsletters.

Mogge, D. *Directory of Electronic Journals, Newsletters, and Academic Discussion Lists.* Washington, DC: Association of Research Libraries, 1997. 950p. ISSN 1057-1337.

A standard reference for serials available via the Internet.

U.S. Department of Commerce Technology Administration. *Directory of Federal Laboratory and Technology Resources: A Guide to Services, Facilities, and Expertise.* Washington, DC: U.S. Department of Commerce Technology Administration, 1993. ISBN 0-934-21340-2.

A listing of federally funded programs provided by National Technical Information Service and National Technology Transfer Center and arranged by subject area. Information includes organization name, institution, address, phone, name and title of contact, facilities or activities, and fields of emphasis.

Alford, J. E., II, ed. *Genetic Engineering and Biotechnology Firms Worldwide Directory.* Princeton Junction, NJ: Mega-Type

Publishing, 703p. $299; alternate format: diskette with custom search and report software—$599.99. *Annual, October.*

International coverage of 6,000 firms with biotech divisions as well as small independent companies. This book is also available on diskette. Information included: company name, address, division name (if applicable), name and title of executives, research lab locations, number of employees engaged in biotech or genetic engineering, equity interests held by others, areas of research activity, and currently available products.

Barry, Inc. *National Biotech Register.* Wilmington, MA: Barry, 1995. 242p. *$58.*

Coverage of 2,200 companies active in biotech research, development, and manufacturing. This book is also available on microfiche. Information included: company name, address, phone, fax, names and titles of key personnel, description of business activities, number of employees, and year founded.

Self, J. *North American Biotechnology Directory.* Houston, TX: I.E.I. Publishing Division, 620p. *Annual. $89.*

This directory lists companies, research institutions, universities, government agencies, suppliers, and manufacturers involved with biotechnology throughout the United States, Canada, and Mexico. This book is also available on diskette and as mailing labels. Information included: company/institution name, address, contacts, phone, and FAX.

Ratafia, M., ed. *TMG's Worldwide Biotechnology and Pharmaceutical Desk Reference.* New Haven, CT: Technology Management Group, 1,311p. *Annual. $372 (book); alternate format: diskette $338 or $398 with print edition included.*

The listing covers more than 375 biotech and pharmaceutical companies, 321 universities, 128 U.S. government agencies and labs, and 402 company funding sources.

This book is also available on diskette. Information included: company/institution name, address, phone, fax, and name and title of contact.

Crafts-Lighty, A., E. Burak Reed, and J. Sime. *UK Biotechnology Handbook.* Slough, England: BioCommerce Data, 620p. *Every 18* *months. $180 + $30 airmail shipping.* Online vendors: DIALOG (Knight-Ridder Information, Inc.), as BioCommerce Abstracts and Directory DataStar (Knight-Ridder Information, Inc.)

Covers more than 700 biotech companies, research institutes, universities, and related professional and academic associations. This book is also available on diskette and as mailing labels. Information included: organization name, phone, telex, fax, names and titles of key personnel, number of employees, description of activities, and areas of research interest.

McGrath, K. A., ed. *Who's Who in Technology.* Detroit, MI: Gale Research, 1,701p. *Irregular. $195.*

This book contains information about more than 25,000 North American men and women working in more than 1,000 scientific and technology fields. Also available as an online database through LEXIS-NEXIS, GALBIO; Vendor: ORBIT Search Service, WHOTECH. Information included: name, occupation, personal data, educational background, career information, technical achievements, organizational affiliations, honors, awards, special achievements, area(s) of expertise, technical publications, patent data, and addresses.

Selected Nonprint Resources

This section provides genetics, biotechnology, and genetic engineering resources available outside of the printed page. Educational video and audiocassettes are reviewed, many of which are designed as instructional aids, others of which are public lectures by leaders in the field or had a first life on educational television. Many can be obtained through Amazon.com or other book dealers.

Electronic handling of information has expanded with the increasing availability of powerful personal computers and the Internet. With computers come more varied presentations of material and the ability to search for specific information. Examples of computer educational software and CD-ROM and DVD contributions enhanced with pictures, text, and audio are provided. Also included are electronic databases and directories of information and sources on aspects of biotechnology and genetic engineering, which are becoming increasingly popular because they are readily expanded and updated and easy to use with powerful search functions. The World Wide Web computer network, accessible from home computers and from computers in many schools and public libraries, is a fertile source for all types of information relevant to genetic engineering ranging from the latest technical nuances to government documents, to personal opinions on issues.

Meaningful use of this information requires that the consumer consider the source carefully because standards for verification of posted facts differ among Web sites. Only a sampling of Web sites on general biotechnology is provided in this chapter because individual Web addresses come and go. Searching for keywords such as *biotechnology, genetic engineering,* or *recombinant DNA* will quickly generate a plethora of interesting sites.

Videocassettes

Biotechnology on Earth
Type: NTSC VHS (U.S. and Canada only)
Age: High school/college
Cost: $58.95
Date: 2001
Source: Hawkhill Associates
ASIN: B00080KVKG

This video attempts to project the impact of the biotechnology and nanotechnology revolutions, the 21st century analogs of the Industrial Revolution that transformed the late 19th and the 20th centuries. The scientific, economic, and ethical issues are outlined from a global perspective, encouraging thinking and discussion rather than dogmatic polemic.

Clone
Type: NTSC VHS (U.S. and Canada only)
Age: 6th grade–college
Length: 60 min.
Cost: $19.95
Date: 2002
Source: National Geographic

Carolina Biological Supply
2700 York Road
Burlington, NC 27215
Phone: (800) 334-5551; FAX: (800) 222-7112

Web site: http://www.carosci.com/
Item: 49-2198

Examines the issues surrounding nuclear transplantation (cloning) and its potential impact on society. Features an interview with Christopher Reeve.

Cloning: How and Why
Type: NTSC VHS (U.S. and Canada only)
Age: High school/college
Cost: $58.95
Date: 2002
Source: Hawkhill Associates
ASIN: B0006UCCW8

This video visits the laboratories in Scotland where the nuclear transfer procedure was used to create the first artificially cloned sheep, Dolly. The procedure is explained by Dr. Neal First in his laboratory at the University of Wisconsin–Madison. The implications of cloning for the 21st century are discussed.

Discovery Channel Bioethics Video Set
Type: VHS video (U.S. and Canada only), set of three, closed-captioned
Age: Grades 6–12
Length: 128 min. total (3 cassettes)
Cost: $129.95
Date: 2002
Carolina Biological Supply
Item: 49-2239

Vol. 1, *Genes and Environment,* Vol. 2, *Genes and Personality,* Vol. 3, *Genes and Body Type.* Describes the relationship between our genes and the environment and how our genetic makeup influences or does not influence individuals, the ongoing nature-versus-nurture continuum.

DNA: The Secret of Life
Type: VHS video/DVD (U.S. and Canada only)
Age: High school and college
Length: 32 min.
Cost: $19.95
Date: 2003

Carolina Biological Supply
Item: 49-0986 VHS; 49-0987 DVD

Chronicles the discovery of the double helical structure of DNA and illustrates the mechanics of genetic engineering. It explores the implications of genetics and the impact of the Human Genome Project.

Double Helix
Type: VHS video (U.S. and Canada only)
Age: Grade 9–college
Length: 108 min.
Cost: $149.95
Date: 1998
Source: BBC Productions
Carolina Biological Supply
Item: 49-2211

A dramatization of the race to solve the structure of DNA. Highlights the problem-solving process that resulted in the unraveling of DNA's double helical organization.

Genetics and Heredity: The Blueprint of Life
Type: VHS (U.S. and Canada only)
Age: Grade 9–college
Length: 22 min.
Cost: $95
Carolina Biological Supply
Item: 49-1063

Describes the field of genetics, which revealed much about probabilities and inheritance before the molecular basis of genes and DNA was discovered. Makes the connection between the modern techniques of molecular biology and genetic engineering and the descriptive power of genetics in explaining the basis of disease.

Genetic Engineering: Exploring the Issues
Type: VHS (U.S. and Canada only)
Age: Grades 7–12
Length: 26 min.
Cost: $ 49.95
Date: 2000
Carolina Biological Supply

Item: 49-2214

Explores the issues that surfaced with the advent of genetic engineering with Dr. Gael Jennings. Should genetic engineering be applied to humans? Who should make that decision, and what impact would it have on society?

Genetic Engineering of Animals
Type: DVD (U.S. and Canada only)
Age: High school and college
Length: 50 min.
Cost: $21.95
Date: 2002
Source: Global Science Productions, Dr. Elliot Haimoff, director
ASIN: B0007VNPEA

HuGEM (Human Genome Education Model)
Tape 1. *An Overview of the Human Genome Project and Its Ethical, Legal, and Social Issues*
Tape 2. *Opportunities and Challenges of the Human Genome Project* (Francis Collins, narrator)
Tape 3. *Issues of Genetic Privacy and Discrimination*
Tape 4. *Genetic Testing across the Lifespan*
Tape 5. *Working Together to Improve Genetic Services*
Type: VHS
Age: Unspecified, for general public
Length: set of 5 videotapes ranging from 19 to 45 min.
Cost: $50 for the set or $15 each, includes a manual
Date: 1997
Source: Georgetown University Child Development Center
Washington, DC 20057
e-mail: Laphamy@medlib.georgetown.edu

or

Alliance of Genetic Support Groups
35 Wisconsin Circle, Suite 440
Chevy Chase, MD 20815
Phone: (301) 652-2553; (800) 336-4363
FAX: (301) 654-0171
e-mail: alliance@cpacess.org

The set of videos is designed to educate the general public. It consists of a series of interviews with researchers, physicians, and consumers.

On Becoming a Scientist
McDougle, V. T. M.
Type: videotape, VHS and teacher's guide
Age: Geared for young audience (unspecified age)
Length: 19 min.
Cost: $70
Date: 1996
Source: Cold Spring Harbor Laboratory Press
10 Skyline Drive
Plainview, NY 11803-2500
Phone: (800) 843-4388; FAX: (516) 349-1946
e-mail: cshpress@cshl.org; Web site: http://www.cshl.org/
ISBN 0-879-69486-6

This video portrays a day in the life of three graduate students and a laboratory manager through a series of interviews. The purpose is to dispel stereotypes and to provide role models.

Stem Cells
Type: VHS (U.S. and Canada only) (closed-captioned)
Age: Grades 9–12
Length: 24 min.
Cost: $86.95
Date: 2002
Carolina Biological Supply
Item: 49-2233

Takes students inside Dr. James Thomson's laboratory where embryonic stem cells were first cultured and explains what stem cells are, how they are handled, and their importance for scientific understanding of disease as well as potential for new therapies. The ethical issues of the production and use of stem cells are also emphasized through interviews with ethicists and scientists.

Computer Programs and Other Electronic Resources

(See also online versions of Directories and Newsletters under **Print Resources**)

If not otherwise noted, these items, and similar materials, are available on Amazon.com and other online vendors.

Biotechnology for Plants, Animals, and the Environment
Type: DVD (U.S. and Canada only)
Age: High school/college
Length: 50 min.
Cost: 21-2217 Teacher's Guide $46.90
21-2208 Student Guide $10.00

Topics covered include agricultural biotechnology, biotechnology of plants, biotechnology of animals, and biotechnology of the environment which involves bioremediation.

PROGRAM: **DNA from the Beginning**
Type: CD-ROM, set of three
Computer: Macintosh/Windows
Age: High school to college
Cost: $54.50
Source: Dolan DNA Learning Center of the Cold Spring Harbor Laboratory

Carolina Biological Supply
2700 York Road
Burlington, NC 27215
Phone: (800) 334-5551; FAX: (800) 222-7112
Web site: http://www.carosci.com/
Item: 21-2200

Explores how genetic engineering was developed and how it works through animations, video interviews with over 80 key scientists, and interactive problems. 72 hours of content.

PROGRAM: **Genetic Engineering Laboratory**
Computer: Macintosh/Windows
Age: Grade 10 to college
Cost: $152.95

Carolina Biological Supply
2700 York Road

Burlington, NC 27215
Phone: (800) 334-5551; FAX: (800) 222-7112
Web site: http://www.carosci.com/
Item: 39-9104

An interactive program that takes students through the molecular biological manipulations of nucleic acids. Experiments can be simulated step by step, from extraction and purification of DNA, through restriction enzyme digests, gel electrophoresis, cloning, amplification, and gene sequencing. This feature is particularly useful for schools that do not have the equipment to do live experiments. A CD-ROM is included (*Genetic Engineering Principles*) to provide students with background in the fundamentals of genetic engineering.

Occupational Outlook Handbook
Type: CD-ROM
Computer: Macintosh/Windows
Age: High school
Cost: $399.00
Carolina Biological Supply
2700 York Road
Burlington, NC 27215
Phone: (800) 334-5551; FAX: (800) 222-7112
Web site: http://www.carosci.com/
Item: W1-39-4996

After an interactive session in which the software surveys the student's educational interests and personality traits, the program provides information from more than 300 career titles on necessary skills, high school courses, length of training, common outlook, related work activities, related occupations, and where to go for additional information. The CD-ROM format includes short video clips for groups of careers, giving the students a visual feel for the work environment. Regular upgrades of new career fields are planned.

Women in Science
Type: CD-ROM
Computer: Macintosh/Windows CD-ROM
Age: Grades 5 through 12
Cost: $79.95
Carolina Biological Supply

2700 York Road
Burlington, NC 27215
Phone: (800) 334-5551; FAX: (800) 222-7112
Web site: http://www.carosci.com/
Item: W1-39-83316

The advantages of the interactive CD-ROM format are evident in the compelling stories of current women scientists and their work. Twelve interview questions are available to hear about these women's lives and their love for science as the student "visits" them where they work. Interactive experiments allow the student to join scientific teams collecting data and analyzing results. A database of information on 130 past and present women scientists is also included.

Databases

Crafts-Lighty, A., ed. **BioCommerce Abstracts and Directory**. Slough, England: BioCommerce Data Ltd. The online vendor is DIALOG (Knight-Ridder Information, Inc. Vendor: DataStar) Knight-Ridder information, Ltd.; Alternate format on diskette.

This database contains 135,000 abstract records on 2,000 U.S. and European biotech companies, research institutes, universities, and related professional and academic associations. Information includes: name, occupation, personal data, educational background, career information, technical achievements, organizational affiliations, honors, awards, special achievements, area(s) of expertise, technical publications, patent data, addresses, production/ services, and research areas. *Semimonthly.*

Specialized Information Services Division, U.S. National Library of Medicine. **Directory of Biotechnology Information Resources.** Bethesda, MD: U.S. National Library of Medicine.

This online database contains more than 3,300 records with information on computer bulletin boards and networks, culture collections, specimen banks, biotechnology centers and related organizations, periodicals, directories and monographs, nomenclature reconciliation, and assorted sources of biotechnology information. Information includes title, organization name, address, phone, name and title of contact, facilities or activities, fields of emphasis, language, and limitations of use and availability of resource. *Quarterly.*

National Human Genome Research Institute. **NHGRI Policy and Legislation Database.** NHGRI's Office of Policy, Communications, and Education, Director Alan E. Guttmacher, M.D. http:// www.genome.gov/LegislativeDatabase.

A free searchable database that is part of the Human Genome Project containing links to full-text copies of federal and state laws and statutes, federal legislative materials, and federal administrative and executive material, including regulations, institutional policies, and executive orders. It currently focuses on genetic testing and counseling, insurance and employment discrimination, newborn screening, privacy of genetic information and confidentiality, informed consent, and commercialization and patenting. The database can be searched by keyword, content type, topic, or source and can sort output by date or citation. There are plans to add more categories of content, primarily foreign laws and statutes, foreign policy, treaty and international agreements, and policy material from international organizations.

Woolum, J., ed. **WooBioScan: The Worldwide Biotech Industry Reporting Database.** Phoenix, AZ: Oryx Press. Online vendor is Knowledge Express data systems, updated monthly. Alternate formats: mailing labels, magnetic media, complete database, mailing list.

The database contains more than 1,000 companies doing product research and development in food processing, agriculture, medicine, and other fields in biotechnology. Information includes company name, address, phone, names and titles of key personnel, number of employees (and number of Ph.D.s), date founded, names and descriptions of subsidiaries, names of investors and percentage of investment, name and description of agreements, and contracts. *Annual, with bimonthly supplements. $975/yr including supplements.*

Internet Sources on Genetic Engineering

These addresses are meant to provide an entry to the wealth of information accessible on the Internet. Interconnections with related sites are easily explored by clicking on the built-in links. Government sites are extensively cross-referenced. The search function on a Web browser will quickly locate many more sites.

Access Excellence. San Francisco: Genentech, Inc. http://www .gene.com/ae

A national educational program sponsored by the biotechnology company Genentech billed in the site header as "A Place in Cyberspace for Biology, Teaching, and Learning." The site provides science updates and clips on newsmakers—the researchers, forensic scientists, and people contributing to the ongoing discussion of bioethics, history of recombinant DNA, and the issues. A variety of activities for teachers and their students is suggested.

Agbiotech Online. Blacksburg, VA: Information Systems for Biotechnology, Virginia Tech. http://gophisb.biochem.vt.edu/

Information on agricultural and environmental biotechnology research, product development, regulatory issues, and biosafety.

Bioethics Journals, Networks, and Associations in the U.S. and Canada. http://www.med.upenn.edu/~bioethic/outreach/bio forbegin/organizations.html

Lists of organizations and publications concerned with bioethics and links to other sites.

Bio Online. Vitadata Corporation. http://www.bio.com/

A comprehensive site for information and services related to biotechnology. It includes resources on industry, government, nonprofit special interest groups, research, career information, and education.

Biotechnology—Center for Food Safety and Applied Nutrition. Food and Drug Administration. http://vm.cfsan.fda.gov/~1rd/ biotechm.html

Web site for food safety that includes a component on regulation of genetically modified food products.

BioTechnology Permits Home Page. http://www.aphis.usda .gov/BBEP/BP/links.html.

Lists of links to other biotechnology Web sites, both U.S. government and international. A good source for information on regulations pertaining to agricultural biotechnology. Provides access to a database (1987–present) on applications for release or testing of bioengineered crops.

Blazing a Genetic Trail. Howard Hughes Medical Institute. http://www.hhmi.org/GeneticTrail/

This nontechnical educational site sponsored by the Howard Hughes Medical Institute is organized in terms of a series of well-written illustrated stories such as "How Genetic Disorders Are Inherited" and "Stalking a Lethal Gene." It is designed to educate the general public.

Council for Responsible Genetics. The Council for Responsible Genetics. http://www.gene-watch.org/

Monitors biotechnology's social, ethical, and environmental consequences. Programs include cloning and human genetic manipulation; genetic testing, privacy, and discrimination; biotechnology and agriculture; biowarfare; genetic bill of rights; and other genetic issues. Publishes magazine *GeneWatch* since 1983. Recent topics include conflicts of interest between academic research and commercial enterprise. The organization aims to foster public debate about social, ethical, and environmental implications of genetic technologies.

DNA Forensics. The Human Genome Project. http://www.ornl .gov/sci/techresources/Human_Genome/elsi/forensics.shtml

Information on the use of DNA in forensics, techniques, legislation, ethics, and other topics.

Gene Letter. The Shriver Center. http://www.geneletter.org/mainmenu.htm

An all-inclusive site for information on the Web about scientific and social issues in genetics with links to federal and state genetics agencies and legal statutes. It includes an uncensored chat area and a search engine for locating information elsewhere in the site and on the Web.

Genetics Education Center. The University of Kansas Medical Center. http://www.kumc.edu/gec/

Genetics education for students.

National Biological Information Infrastructure. http://www .nbii.gov/

A main Web site for biological information of all sorts on the Web. There are links to many scientific disciplines and issues, regulations, and so on.

National Human Genome Research Institute. http://www .nhgri.nih.gov/

The main Web site for Human Genome Research. It is the access point for information on the project and for databases derived from it. Includes research on bioethics and other areas impacted by the project.

The Pure Food Campaign. http://interactivism.com/purefood/

Web site for an activist group focusing on issues of food production, organic versus genetically engineered or processed. The Pure Food Campaign is allied with Jeremy Rifkin's Foundation on Economic Trends. Links are available to other similar types of groups.

United States Department of Agriculture. http://www.ice.net/ jumps/ag/ag.html

Includes many links to agricultural biotechnology sites.

United States Regulatory Agencies Unified Biotechnology. http://usbiotechreg.nbii.gov/lawsregsguidance.asp

U.S. laws and regulations; agency responsibilities.

Gaulin, P. **2001 Web Site Source Book: A Guide to Major U.S. Businesses, Organizations, Agencies, Institutions, and Other Information Resources on the World Wide Web.** Detroit, MI: Omnigraphics, 2001. 2,536p. ISBN 0-780-80428-7.

Appendix:
Acronyms

AAV	Adeno-associated virus
ADA	Adenosine deaminase—an enzyme involved in nucleotide metabolism
AIDS	Acquired immunodeficiency syndrome
ALS	Amyotrophic lateral sclerosis—a neurodegenerative disease
APHIS	Animal and Plant Health Inspection Service
BIO	Biotechnology Industry Organization
BSCC	Biological Science Coordinating Committee
Bt	*Bacillus thurengensis* toxin
CEO	Chief executive officer
CF	Cystic fibrosis
CODIS	Combined DNA Index System
CRADA	Cooperative Research and Development Agreement
DOE	Department of Energy (U.S.)
ELSI	Ethical, Legal, and Societal Implications—Task Force of the Human Genome Project
EPA	Environmental Protection Agency
EST	Expressed Sequence Tag
FASEB	Federation of American Societies for Experimental Biology
FBI	Federal Bureau of Investigation
FDA	Food and Drug Administration
FSIS	Food Safety and Inspection Service
GATT	General Agreement on Tariffs and Trade
G-CSF	Granulocyte stimulating factor
GMO	Genetically modified organism
GURT	Genetic Use Restriction Technology
HD	Huntington's disease—a neurodegenerative disease affecting muscular coordination
HGS	Human Genome Sciences

HIPAA   Health Insurance Portability and Accountability Act
HIV   Human immunodeficiency virus
IND   Investigational new drug application
NHGRI   National Human Genome Research Institute
NIH   National Institutes of Health
NSF   National Science Foundation
OTA   Office of Technology Assessment
PCB   Polychlorinated biphenyl (complex environmentally stable organic molecule)
PCR   Polymerase chain reaction
PMN   Premanufacture notification
RAC   Recombinant DNA Molecule Program Advisory Committee
RFLP   Restriction fragment length polymorphism
S&E   U.S. Department of Agriculture, Science, and Education
TIGR   The Institute for Genomic Research
U.S.C.   United States Code (legal code)
USDA   United States Department of Agriculture

Glossary

aerobe An organism that can grow in the presence of oxygen.

Amino acid A chemical building block used by cells to make proteins and that can be converted into other chemicals the cell needs. There are 20 natural amino acids used to make proteins.

anaerobe An organism that cannot grow in the presence of oxygen.

antibiotic A chemical produced by one organism that kills or inhibits growth of another organism in competition for food or space. Often produced by microorganisms (yeast or bacteria), humans have adopted their use to control organisms that cause disease.

antibodies Defense proteins produced by the body in response to either vaccines or the organism or a component that binds and neutralizes the toxic agent, allowing the body to clear the danger.

aquifer A natural underground collection of water accumulated in a layer of permeable rock or sand from percolating surface water usually cleansed by its passage through soil layers. Polluted soils are leading to contamination of the underground water, which normally has a very slow turnover. This pool is often the source of fresh water for human consumption or for agriculture.

aromatic Organic chemical compound with a very low hydrogen-to-carbon ratio. Because of their resistance to biological modification, aromatic compounds can be quite stable in the environment and are often considered pollutants. Examples are benzene, phenol, naphthylene, dioxin, 2,4-D, and biphenyls.

bacteria (bacterium, singular) Single cell microorganisms capable of multiplying independently, the genetic information of which is not enclosed in a membrane-bounded nucleus. Bacteria are thought to be one of the oldest forms of life.

Bacteriophage (also phage) A virus that infects bacteria. Used by molecular biologists as a vehicle to transfer DNA sequences into host bacteria for study.

base pair Technical term that refers to the nucleotides paired across the double-stranded DNA helix by hydrogen bonds (A=T; C=G). One nucleotide in a single DNA (or RNA) sequence corresponds to one base pair in a double-stranded DNA (or RNA) sequence.

biodegradable Capable of being broken down into its component parts by biological organisms.

bioethics Application of the ideas of ethical conduct to biotechnology and medicine. Trying to reconcile the ability to accomplish certain medical or genetic or environmental actions with the societal norms of acceptable actions. Being able to perform a genetic test does not mean that it should be done, especially if there is no cure for the disease. Some people consider doing a genetic test for an incurable disease once a person is born to be unethical.

biomass Total dry weight of biological organisms.

bioremediation The use of biological organisms to remove toxins and wastes from contaminated materials.

candidate gene In gene mapping, a gene selected for study as being potentially responsible for a trait based on an activity or property of the molecule encoded by the nucleotide sequence.

catalyst An agent that speeds up the rate of a chemical reaction. In cells, catalysts are special proteins called enzymes that allow reactions to occur at body temperature.

central dogma A hypothesis stating that the direction of genetic information flow is from DNA through RNA (messenger RNA) into protein. This turns out to hold for most situations except notably for certain viruses that contain RNA as their genetic material. These are called retroviruses, referring to their ability to replicate their RNA through a DNA intermediate that they generate by the action of a reverse transcriptase enzyme.

chromosome Structures in the nuclei of cells that contain the linear sequence of genes on DNA wrapped for compactness on a protein scaffold. A gene resides in a particular position on a particular chromosome in all cells of an individual in a species. Human cells contain 23 pairs of chromosomes. Reproductive cells contain only one copy of each chromosome so that when the sperm and ovum combine, the resulting zygotic cell (soon to become an embryo) contains one set of chromosomes (and genes) from each parent.

clone A genetic replica, either an organism or a piece of DNA.

codon A group of three consecutive nucleotides in the DNA polymer (64 possibilities) that codes either for one of the 20 amino acids or peptide chain termination.

cytoplasm The semisolid fluid that fills plant, animal, and microbial cells and in which are immersed the nuclear material and other cellular structures that carry out the chemistry of life.

deoxyribonucleic acid (DNA) A chain or polymer that contains the genetic information for an organism. It comprises four kinds of units called nucleotides: adenosine (A), thymidine (T), cytosine (C), and guanosine (G).

DNA ligase An enzyme that chemically joins separated (usually restriction enzyme-cleaved) double-stranded DNA segments together end to end to produce a continuous DNA molecule. Another important tool of recombinant DNA technology.

dominant gene A genetic characteristic that is expressed when present in either of the set of chromosomes from the parents carrying that DNA sequence. In genes controlling eye color, brown eyes are a dominant characteristic.

ecology The study of the interaction of groups of organisms with each other and with their environment.

ecosystem A unit consisting of a natural community of organisms together with their environment.

enzyme Proteins that speed up (catalyze) a chemical reaction and control the generation of products.

ethics A philosophy or system for judging right and wrong actions in human conduct. Commonly termed *morality,* the basic assumptions about what constitutes right and wrong are often difficult to agree on within any particular society. Complicating the issue is that these basic assumptions can vary dramatically between societies with different histories. Because they are usually deeply held convictions rather than verifiable facts, rational discussion and compromise are difficult.

eukaryote A cell in which DNA is contained within a membrane-bounded nucleus. These include such single-celled organisms as yeast and fungi. Cells other than bacteria or blue-green algae are generally eukaryotic. Viruses are not cells and are not eukaryotic.

F1 hybrid The first offspring from a genetic cross between two organisms with different genetic backgrounds that have an equal contribution of genetic material from each parent. F0 is the parental generation. Crossing two F1 generation organisms will produce a ratio of two parental and two hybrids in the second (F2) generation. Hybrid qualities are rapidly diluted out in the general population if the seeds produced by an F1 crop are replanted by the farmer. Providing only F1 seeds keeps the farmer dependent on the supplier for new seed.

fermentation Originally restricted to refer to the growth of microorganisms utilizing organic compounds as a source of energy in the absence of atmospheric oxygen, the term is now used to refer to the controlled culture of microorganisms to produce useful products.

forensic Suitable for courts of law or for public debate.

gene A segment of DNA with sequence codes for a specific function or protein molecule. The gene is the basic unit of genetic information and can extend for millions of nucleotides in length.

gene chip A microfabricated system containing an array of gene nucleotide sequences used for determining expression levels of messenger RNAs in cells or tissues. Tens of thousands of genes can be displayed on a single chip.

gene linkage The tendency of genes to be inherited together on the same piece of DNA.

gene mapping A process of matching inheritance of a trait with a location on a chromosome to identify the gene involved.

gene therapy A therapeutic approach in molecular medicine in which the genetic code is altered in the patient or additional genetic coding is added where there is a deficient gene. The modification can be limited to certain cells and not be transmitted to the next generation (somatic therapy). Alternatively, the genetic information of the reproductive cells can be modified, with that change carried through the offspring of the patient into subsequent generations (germ line therapy).

genome All nuclear genetic material in an organism, including genes and intervening nucleic acid sequences in all of the chromosomes of an organism.

germ cells The cells responsible for transmitting the heredity information during reproduction. In mammals these are sperm (male) and ova (female).

germplasm A general term for the living material that controls heredity in a species. It refers to the genetic potential of the species.

Green Revolution Intensive agriculture aimed at increasing crop yields to allow developing nations to become self-sufficient in food production through the use of special high-yield hybrid disease-resistant plant strains and increased use of fertilizers, herbicides, and irrigation. Although these goals were attained, analysis suggests that considerable damage to local ecosystems occurred through overuse of chemical supplements. In addition, residual economic dependency for the fertilizer and herbicides casts a shadow over plans being made to introduce recombinantly modified crops.

GURT Genetic use restriction technology—a controversial use of genetic engineering designed to control expression of designer characteristics to ensure the supplier a market. Currently applied to crops genetically modified to be insect- or herbicide-resistant to limit reuse of seeds or require additional technology to be purchased by the grower to express the desirable trait.

informatics The technology of information management and processing. Bioinformatics involves the manipulation of data on nucleotide sequences and all of the associated information.

insert A familiar word used to refer to the piece of DNA placed or "inserted" into a plasmid used as a genetic engineering tool.

in silico Operations carried out by computer analysis without physical experiments, literally in silicon, referring to the computer semiconductor chips.

membrane Cellular membranes comprise two layers of phospholipids, molecules with a phosphate head group that faces the water and with a greasy hydrocarbon tail. The tails contact each other, forming an oil-like layer to provide a waterproof barrier between the outside and inside of the cell to isolate the cellular chemistry inside from outside influences. Proteins with special functions to control transport of vital molecules from the environment are embedded in the membrane.

metabolism The cellular chemical processes that provide energy and material for cellular function. A metabolite is any chemical entity involved in these chemical processes.

microbe Microscopically small, single-celled organisms, usually used to refer to bacteria, algae, and yeast.

microscope An instrument used to make images of small objects. A **light microscope** uses glass lenses to bend visible light passing through a thin sample to provide magnification of up to 2000X. An **electron microscope** uses magnetic lenses to focus an electron beam, magnifying up to 1 million fold.

mitochondrion A part of the cell, an **organelle,** roughly the size of a bacterium, that contains the biochemical machinery to provide the chemical energy needed to run a cell.

monocotyledon One of a large family of flowering plants that include the grasses that have only one leaf in the newly emerged seedling and usually parallel veins in the leaves. Many cereal grains are monocots, which have presented some difficulties in genetic engineering, and hence these crops have lagged in development.

multigenic trait A genetic characteristic governed by several genes that contribute to varying extents. Many chronic diseases such as heart disease or certain types of behaviors or inherited personality types are multigenic. Drought resistance and cold hardiness are multigenic traits in plants.

niche An ecological term that refers to the specific position in an ecosystem to which its occupant is adapted.

nuclear transfer Replacing the nucleus and its associated genetic material of one cell with the nucleus of another cell. Erroneously dubbed "cloning" by the media, nuclear transfer performed with sheep and with other species has produced an identical genetic copy of the nucleus donor, except for cytoplasmic genetic material in the mitochondria. Although the transplantation was not true cloning, the feat resurrected the controversy over possible human genetic engineering and led to a raft of pending legislation to curb such experimentation.

nucleotide The building block of DNA and RNA. Each nucleotide comprises one of four chemical "bases" (adenine, cytosine, guanine, or thymine [uracil in RNA]), one ribose sugar, and a phosphate group.

nucleus Compartment of a cell containing the genetic material. In animal and plant cells, this region is enclosed by a membrane and contains the chromosomes.

organelle A small part of a cell organized to perform a specialized function. The **nucleus** stores the genetic material and controls information flow from the DNA to the rest of the cell. The **ribosome** assembles proteins from information provided by the nucleus through **messenger RNA.** The **mitochondrion** provides cellular energy and synthesizes needed **metabolites**.

organic The scientific definition is a chemical compound containing carbon and hydrogen. A popular definition refers to food or products produced without the aid of purified chemicals for nutrients or processing, often relying on complex biological sources for those materials.

pathogen Any organism capable of causing disease.

PCR Polymerase chain reaction—A molecular biological technique using DNA replicating enzymes (polymerases) isolated from bacteria living in hot springs or deep ocean hydrothermal vents. This revolutionary method employs cycles at high temperature to replicate or amplify specific nucleic acid sequences rapidly from a complex mixture of sequences. Various modifications allow the technology to perform other reactions using the polymerases to alter target gene sequences.

pharmacogenomics Use of genomics to personalize pharmaceutical applications according to genetic propensity to respond to a particular therapeutic regimen. This attempts to take into account individual genetic variability in metabolizing drugs or in responding to a drug based on genetic markers linked to these variations.

phytoremediation Application of the normal growth and uptake properties of plants for the removal of toxic agents from the environment by concentration or by metabolism.

plasmid Pieces of DNA-containing genes, usually circular, that can be transferred between microorganisms and into cells and that cause themselves to be copied by the recipient cell. Because they can be handled so easily, they are a favorite tool of molecular biologists for storing inserted pieces of DNA and causing the added gene to be activated under experimental control.

pluripotent A term used in referring to stem cells that are capable of differentiating down a particular path, giving rise to a number of, but not all, types of cells. Compare **totipotent.**

polymer A chain formed from similar types of building blocks. Typical biological building blocks and their common polymers are amino acids (proteins), carbohydrates (cellulose), and nucleic acids (DNA and RNA). Nonbiologic polymers include tars, waxes, polyesters, and polyamides (nylon, plastics), although genetic engineering is allowing plants to carry out some of these syntheses.

polymorphism A single nucleotide change in the sequence of a gene in a species that is present in a significant percentage of a population (as opposed to a mutation that alters one organism). The change may or may not result in a change of function of the nucleotide sequence or in a protein, but its inheritance can be followed through generations as a marker.

prokaryote A cell such as a bacterium in which DNA is not contained within a membrane-bounded nucleus. The DNA usually is found in a single long, circular strand. Compare **eukaryote.**

protein An unbranched chain or polymer of 20 types of units called amino acids. Proteins provide structure and catalytic activity for cells to synthesize and organize the rest of their components.

recessive gene A genetic characteristic that is only evident when both pairs of chromosomes carry that particular DNA sequence. In genes controlling human eye color, blue eyes (in adults) are a recessive characteristic. Compare **dominant gene.**

recombinant DNA The manipulation of DNA and genes in new combinations. Although this process also takes place in nature, this term is normally used to refer to human-mediated rearrangements.

replication The copying of the DNA sequence of the **genome.**

restriction enzyme An enzyme (endonuclease) capable of recognizing a specific DNA sequence of four or more base pairs and cleaving the double-stranded helix. Evolved by bacteria to prevent incorporation of foreign DNA into their genomes in sort of a genetic immune system, molecular biologists use the isolated enzymes to cut and paste DNA segments into the desired order. The ability to manipulate DNA combinations at will was the key to the recombinant DNA revolution and genetic engineering.

reverse transcriptase An enzyme produced by viruses in which the genetic material is RNA rather than the usual DNA. It possesses the unusual property of copying RNA sequences into DNA (an RNA-dependent DNA polymerase). This property has been exploited by molecular biologists who use the enzyme to copy messenger RNA into a complementary DNA sequence that can then be replicated and engineered in the standard way in bacteria and other organisms.

ribosome A part of the cell, an **organelle,** that organizes the connecting of amino acids together to produce a protein.

ribozyme A special RNA species with catalytic activity (usually RNA-cleaving activity). These molecules can participate in splicing of RNA messages and in the destruction of specific RNA sequences. Synthetic ribozymes can be constructed to control the levels of specific messages in cells.

species A group of like organisms classified together. Individuals within a species can interbreed. Different species cannot interbreed to produce fertile offspring.

statistical genetics A branch of genetic analysis that uses statistical techniques to associate observed traits with specific stretches of DNA. It is used to map the location of genes on chromosomes.

stem cell A primordial cell that has not yet undergone the developmental process to take on the specialized characteristics of a tissue type or organ. Differentiation is a multistep, multipath process and stem cells can arrest at various points along the path awaiting a signal to form a particular kind of cell.

strain A variant within a species. Intraspecific strains can interbreed.

totipotent A term used in referring to stem cells that are capable of differentiating into any type of cell. The final product cell type is determined by specific biological factors and cellular environment as well as the order in which the cells are exposed to the influences. Embryonic stem cells and germinal stem cells are totipotent.

transcription factor A protein that regulates the activity of genes. It binds to a specific DNA sequence in the regulatory part of a gene and organizes several proteins and the enzyme needed to synthesize messenger RNA.

transgene A gene from another species of organism. A mouse engineered to contain the human hemoglobin gene is a transgenic animal. These are artificial combinations of genes that would not have occurred with any appreciable frequency in the natural environment.

transgenic organism An organism containing genetic material from another species.

tumorigenic Capable of causing a cell to divide out of control and to become cancerous.

vaccine A preparation of killed or weakened pathological organisms or parts of those organisms administered to induce immunity against that pathogen.

vector A piece of DNA initially borrowed from a naturally occurring bacterial **plasmid** or from a **virus** and engineered to contain convenient features for molecular biologists. Vectors allow the easy handling of pieces of DNA, allowing the replication of the DNA, synthesis of RNA, or production of protein from a cloned gene in whatever type of host cell the vector is designed to function.

virus A small protein and lipid particle containing DNA or RNA sequences that can penetrate into cells and direct their own replication. Parasitic, lacking metabolic functions, and unable to replicate themselves without assistance, viruses are generally not considered to be alive.

xenograft Transplantation of cells or tissue from one species into another such as replacement of a human heart with a baboon heart.

Index

About the Author

Harry LeVine, III, Ph.D., received his B.S. in biochemistry from Cornell University in 1971 and his Ph.D. in physiological chemistry from the Johns Hopkins School of Medicine in 1975. A postdoctoral stint at the Burroughs-Wellcome Company evolved into a 28-year career in the pharmaceutical industry encompassing tenures at Wellcome, Glaxo, Parke-Davis, and Pfizer. His first experience of the recombinant DNA revolution was wondering about the relevance of a qualifying exam question asked by two professors on a then-esoteric bacterial DNA restriction enzyme system. Ham Smith and Dan Nathans subsequently won the Nobel Prize for their work in that area. Later, Dr. LeVine's projects in multiple therapeutic areas depended heavily on DNA technology and then expanded into new fields in the era of the genome. In 2003, he moved to the University of Kentucky where he is associate professor of biochemistry in the Department of Molecular and Cellular Biochemistry and the Sanders-Brown Center on Aging, where he is studying Alzheimer's disease. Dr. LeVine is author of more than 100 research articles, reviews, and book chapters as well as four patents. In addition to *Genetic Engineering: A Reference Handbook* (ABC-CLIO), he is coauthor of *Drug Discovery and Development—Technology in Transition* (Churchill/Livingston, 2006). He also has published articles about science and scientists for young readers in *Cricket* and *Cicada* magazines.